Commutative Matrices

ACADEMIC PAPERBACKS*

EDITED BY Henry Booker, D. Allan Bromley, Nicholas DeClaris, W. Magnus, Alvin Nason, and A. Shenitzer

---

BIOLOGY

Design and Function at the Threshold of Life: The Viruses
   HEINZ FRAENKEL-CONRAT
The Evolution of Genetics   ARNOLD W. RAVIN
Isotopes in Biology   GEORGE WOLF
Life: Its Nature, Origin, and Development   A. I. OPARIN
Time, Cells, and Aging   BERNARD L. STREHLER
The Spread of Cancer   JOSEPH LEIGHTON

ENGINEERING

A Vector Approach to Oscillations   HENRY BOOKER
Dynamic Programming and Modern Control Theory   RICHARD
   BELLMAN and ROBERT KALABA
Hamilton's Principle and Physical Systems   B. R. GOSSICK

MATHEMATICS

Finite Permutation Groups   HELMUT WIELANDT
Complex Numbers in Geometry   I. M. YAGLOM
Elements of Abstract Harmonic Analysis   GEORGE BACHMAN
Geometric Transformations (in two volumes)   P. S. MODENOV
   and A. S. PARKHOMENKO
Introduction to p-Adic Numbers and Valuation Theory
   GEORGE BACHMAN
Linear Operators in Hilbert Space   WERNER SCHMEIDLER
The Method of Averaging Functional Corrections: Theory and
   Applications   A. Yu. LUCHKA
Noneuclidean Geometry   HERBERT MESCHKOWSKI
Quadratic Forms and Matrices   N. V. YEFIMOV
Representation Theory of Finite Groups   MARTIN BURROW
Stories about Sets   N. Ya. VILENKIN
Commutative Matrices   D. A. SUPRUNENKO and
   R. I. TYSHKEVICH

PHYSICS

Crystals: Their Role in Nature and in Science   CHARLES BUNN
Elementary Dynamics of Particles   H. W. HARKNESS
Elementary Plane Rigid Dynamics   H. W. HARKNESS
Mössbauer Effect: Principles and Applications
   GUNTHER K. WERTHEIM
Potential Barriers in Semiconductors   B. R. GOSSICK
Principles of Vector Analysis   JERRY B. MARION

*Most of these volumes are also available in a cloth bound edition.

Commutative Matrices

D. A. SUPRUNENKO

and

R. I. TYSHKEVICH

Translated by Scripta Technica, Inc.

1968

ACADEMIC PRESS *New York and London*

169858

Originally published in 1966 by Nauka and Tekhnika press,
Minsk, U.S.S.R., in the Russian language under the title
"Perestanovochnyye Matritsy."

ACADEMIC PRESS INC.
111 Fifth Avenue, New York, New York 10003

United Kingdom Edition published by
ACADEMIC PRESS INC. (LONDON) LTD.
Berkeley Square House, London W.1

Library of Congress Catalog Card Number: 68–18683

Printed in the United States of America

Foreword

In this book the authors deal with elementary properties of a system of commutative matrices, with general properties of commutative matrix algebras over an arbitrary field, and with certain classification questions pertaining to the theory of maximal commutative subalgebras of the full matrix algebra over the field of complex numbers. In addition, the authors present a number of unsolved problems in the area of commutative matrix algebras.

The book is intended for scientists and students of physics and mathematics who are interested in the calculus of matrices.

Contents

1

Elementary Properties of Commutative Matrices

Let P denote an arbitrary field, let P^n denote an n-dimensional linear space over the field P, let P_n denote the algebra of all $n \times n$ matrices over the field P or the algebra (isomorphic to it) of all linear operators defined on the space P^n, and let $GL(n, P)$ denote the group of all nonsingular $n \times n$ matrices over the field P or the group of all nonsingular linear operators defined on the space P^n. Two matrices (or two operators) a and b in P_n are said to be *commutative* (or to *commute*) if $ab = ba$. In the present chapter we expound the simplest properties of commutative matrices.

If L is a linear space over the field P, then we denote by $L:P$ the dimension of L over P. For $u_1, \ldots, u_k \in L$, we shall denote by $[u_1, \ldots, u_k]$ the subspace generated by the set $\{u_1, \ldots, u_k\}$. We shall use the same symbol $[A_1, \ldots, A_k]$ to denote a quasi-diagonal matrix with blocks $A_1, \ldots, A_k$ on the principal diagonal.

1

1. Schur's Lemma

Let f denote a linear operator defined on the space P^n and let Q denote a subspace of P^n. Then f defines a mapping φ of Q into P^n: specifically, $\varphi q = fq$ for $q \in Q$. The mapping φ is called the *restriction* of f to Q and is denoted by $f|Q$. Now, if Σ is a system of operators in P_n, the set of all $\sigma|Q$, where $\sigma \in \Sigma$, is called the restriction of Σ to Q and is denoted by $\Sigma|Q$.

Let Σ denote a system of linear operators in P_n. If Q is a subspace of the space P^n such that, for arbitrary $q \in Q$ and $\sigma \in \Sigma$, we have $\sigma q \in Q$, then Q is called an *invariant subspace* with respect to Σ. A system of linear operators Σ is said to be *irreducible* if P^n contains only two subspaces that are invariant with respect to Σ, namely, the subspace $Q = (0)$, that is, the subspace consisting only of the zero vector, and $Q = P^n$. Otherwise, Σ is said to be a *reducible system* of operators.

Let Q denote a subspace that is invariant with respect to Σ. If $\Sigma|Q$ is an irreducible system, we shall say that Q is an *irreducible subspace* with respect to Σ. If the space P^n is the direct sum of subspaces that are irreducible with respect to Σ, we say that Σ is *completely reducible*.

We state without proof the following fact, which is familiar from linear algebra:

Proposition 1.   *Let Σ denote a completely reducible set of transformations in P_n and let Q denote a subspace, other than (0) and P^n, that is invariant with respect to Σ. Then, P^n contains a subspace R that is invariant with respect to Σ such that*

$$P^n = Q \dotplus R,$$

where the symbol $\dotplus$ denotes the direct sum.

We shall call R the *invariant complement* of the subspace Q.

Now, let Φ denote a system of matrices in P_n, let

$$u_1, u_2, \ldots, u_n \tag{1.1}$$

denote a basis for the space P^n, and let Σ denote the system of linear operators corresponding to Φ in the basis (1.1). The system Φ is said to be *irreducible* (*reducible, completely reducible*) if the system of operators Σ is irreducible (reducible, completely reducible).

Now, let A denote a set. To every $a \in A$, we assign a matrix $N(a)$ in P_n. We shall call such a mapping a *representation* of the set A[1]:

$$a \to N(a). \tag{1.2}$$

Two representations $a \to N(a)$ and $a \to M(a)$ of the set A are said to be *equivalent* if, in the full linear group $GL(n, P)$, there exists a matrix s such that $s^{-1}M(a)s = N(a)$ for all $a \in A$. The representation (1.2) is said to be irreducible if the set of all matrices $N(a)$ is irreducible.

Let Σ denote a completely reducible set of linear operators in P_n. Suppose that

$$P^n = Q_1 \dotplus \cdots \dotplus Q_s \tag{1.3}$$

is a decomposition of the space P^n as the direct sum of invariant subspaces that are irreducible with respect to Σ.

---

[1] The mapping (1.2) is not necessarily a homomorphism since A is an arbitrary set.

Then, the decomposition (1.3) defines s irreducible representations γ_j of the system Σ:

$$\gamma_j : \Sigma \to \Sigma | Q_j, \qquad \gamma_j(\sigma) = \sigma | Q_j, \qquad \sigma \in \Sigma. \tag{1.4}$$

Proposition 2. *Suppose that Σ is a completely reducible set of linear operators in P_n and suppose that $P^n = Q_1 \dotplus \cdots \dotplus Q_s$ and $P^n = R_1 \dotplus \cdots \dotplus R_t$ are two decompositions of the space P^n as the direct sum of irreducible invariant subspaces with respect to Σ. Then, $s = t$ and there exists a numbering of the direct summands R_j such that the representation $\varphi_j : \Sigma \to \Sigma | R_j$, where $\varphi_j(\sigma) = \sigma | R_j$ for $\sigma \in \Sigma$, is equivalent to the representation $\gamma_j : \Sigma \to \Sigma | Q_j$, where $\gamma_j(\sigma) = \sigma | Q_j$ for $j = 1, \ldots, s$.*

Lemma (*Schur*). *Let*

$$a \to N(a) \tag{1.5}$$

and

$$a \to M(a), \qquad a \in A \tag{1.6}$$

denote two irreducible representations of the set A by matrices in P_n and P_m, respectively. If σ is an $m \times n$ matrix over P such that

$$M(a)\sigma = \sigma N(a) \tag{1.7}$$

for all $a \in A$, then either σ is the zero matrix or σ is a nonsingular square matrix and the representations (1.5) and (1.6) are equivalent.

Proof: We shall treat σ as the matrix of a linear operator mapping the space P^n into P^m and, for convenience in notation, we shall not distinguish σ from this operator.

Similarly, we shall not distinguish the matrices $M(a)$ and $N(a)$ from the operators of which they are the matrices. It follows from Eq. (1.7) that

$$M(a)(\sigma P^n) = (M(a)\sigma)P^n = (\sigma N(a))P^n$$

$$= \sigma[N(a)P^n] \subset \sigma P^n,$$

that is, σP^n is an invariant [with respect to the set of all matrices $M(a)$] subspace of the space P^m. But (1.6) is an irreducible representation. Therefore, either $\sigma P^n = (0)$ or $\sigma P^n = P^m$. In the first case, σ is the zero operator. In the second case, the operator σ maps the space P^n homomorphically onto P^m. Suppose that the second case obtains. Let us look at the kernel Q of this homomorphism. It follows from Eq. (1.7) that

$$\sigma[N(a)Q] = [\sigma N(a)]Q = M(a)[\sigma Q] = M(a)(0) = (0).$$

Thus, Q is a subspace of the space P^n that is invariant with respect to the set of all matrices $N(a)$. Since the representation (1.5) is also irreducible, either $Q = (0)$ or $Q = P^n$. If $\sigma \neq 0$, the latter is impossible. Consequently, $Q = (0)$, $m = n$, and σ is a nonsingular operator. Then, it follows from Eq. (1.7) that $\sigma^{-1}M(a)\sigma = N(a)$ for all $a \in A$. This completes the proof of the lemma.

If $M \in P_n$, then the set of all operators in P_n that commute with every operator in the set M is called the *centralizer* of the set M in P_n.

Corollary 1. *If M is an irreducible subset of P_n, then every nonzero operator of the centralizer Z of the set M in P_n is nonsingular; hence, Z is a division ring.*

Proof: If $z \in Z$ and m ranges over M, then $zm = mz$. It then follows from Schur's lemma that either $z = 0$ or z is a nonsingular operator.

Corollary 2. *Suppose that $A \neq (0)$ is an irreducible commutative subalgebra of P_n. Then,*

   (i)   *A is a field;*

   (ii) *the degree of the field A over P is equal to n;*

   (iii) *the centralizer of the algebra A in P_n coincides with A.*

Proof: (i)   In accordance with Corollary 1, every nonzero element of A is a nonsingular operator and hence A is a commutative finite-dimensional algebra without divisors of zero. Thus, A is a field.

(ii)   Let u denote an arbitrary nonzero vector in P^n. Consider the space Au. Obviously, Au is a subspace of P^n that is invariant with respect to A. Since A contains a nonsingular matrix, $Au \neq (0)$ and, consequently, $Au = P^n$. If $w_1, \ldots, w_k$ is a basis of the field A over P, the vectors

$$w_1 u, \ldots, w_k u \tag{1.8}$$

are linearly independent. To see this, suppose that

$$\sum_i \lambda_i (w_i u) = 0,$$

where $\lambda_i \in P$. Then,

$$\left(\sum_i \lambda_i w_i \right) u = 0.$$

Since $\sum_i \lambda_i w_i$ is either zero or a nonsingular operator, we have

$$\sum_i \lambda_i w_i = 0,$$

where $\lambda_i = 0$ for $i = 1, \ldots, k$, so that the system (1.7) is linearly independent. It is now obvious that (1.8) is a basis for the space P^n. Therefore, $k = n$.

(iii)   Suppose now that Z is the centralizer of A in P_n and that $g \in Z$. Consider the algebra B generated by g and A. On the basis of (i) and (ii), B is a field whose degree over P is n. Consequently, $B = A$, $g \in A$, and $Z = A$.

The converse of this corollary is also true:

Lemma 1. *The field of matrices of A containing only the matrix E_n is completely reducible. All representations of the algebra A of the form* (1.4) *that are defined by the decomposition of the space P^n as a direct sum of irreducible subspaces invariant with respect to A are equivalent.*

Proof: Let u denote a nonzero member of P^n. Then Au is an irreducible invariant subspace of P^n with respect to A and $Au \neq (0)$. If $Au = P^n$, the conclusion of the lemma is immediate. If $Au \neq P^n$, let v denote a nonzero member of $P^n \setminus Au$. Then, $Au \dotplus Av$ is a direct sum. Continuing in this way, we see that the space P^n is the direct sum of irreducible invariant subspaces Au_k for $k = 1, \ldots, s$. If $w_1, \ldots, w_m$ is a basis for the field A over P, then $w_1 u, \ldots, w_m u$ is a basis for the space Au and, consequently,

$$w_1 u_1, \ldots, w_m u_1, \ldots, w_1 u_s, \ldots, w_m u_s \qquad (1.9)$$

is a basis for P^n. In the basis (1.9), the matrices of the algebra A are of the form $[b, b, \ldots, b]$, where b ranges over an irreducible field of matrices in P_m.

Corollary 3. *The minimum polynomial of the operator* σ *belonging to the centralizer in* P_n *of an irreducible subset* M *of* P_n *is irreducible over* P.

Proof: If the minimum polynomial $\varphi(x)$ of the operator σ is decomposed over P as a product $\varphi(x) = \varphi_1(x)\varphi_2(x)$ of polynomials of nonzero degree, then $\varphi(\sigma) = \varphi_1(\sigma)\varphi_2(\sigma) = 0$. Consequently, $\varphi_1(\sigma)$ is a singular nonzero operator.   Obviously, $\varphi_1(\sigma)$ commutes with each operator in the set M, which contradicts Schur's lemma.

Corollary 4. *If the field* P *is algebraically closed, then the centralizer in* P_n *of an irreducible set of operators in* P_n *consists of all scalar operators* λE_n, *where* $\lambda \in P$ *and* E_n *is the identity operator.*

This assertion follows from Corollary 3 and the fact that a polynomial of degree greater than one is reducible over an algebraically closed field.

Corollary 5. *If a field* P *is algebraically closed and* $n > 1$, *then* P_n *contains no irreducible set of pairwise-commutative operators.*

Proof: If M is an irreducible set of pairwise-commutative operators in P_n, then, on the basis of Corollary 4, all operators in M are scalar operators.   Therefore, $n = 1$.

Corollary 6. *Let* P *denote an arbitrary field and let* M *denote the set of quasi-diagonal matrices* $g \in P_n$ *of the form* $g = [A_1(g), \ldots, A_s(g)]$, *where* $g \to A_i(g)$ *is an irreducible representation by matrices of degree* n_i, *where* $\sum_i n_i = n$. *If* σ *belongs to the centralizer of the subset* $M \subset P_n$, *then*

$$\sigma = \|a_{ij}\|, \qquad i, j = 1, \ldots, s, \qquad (1.10)$$

where a_{ij} is an $n_i \times n_j$ matrix consisting only of zeros when the representations $g \to A_i(g)$ and $g \to A_j(g)$ are nonequivalent, but is a square nonsingular or zero matrix when these representations are equivalent.

Proof: Let us represent σ in the form (1.10). Since $\sigma g = g \sigma$ for all g in M, we have $a_{ij} A_i(g) = A_j(g) a_{ij}$. The desired result then follows from Schur's lemma.

2. Normal Forms of a System of Commutative Matrices

In what follows, we shall need some elementary information from the theory of associative algebras. We shall state these without proof as applied to the commutative case.

Let A denote an associative commutative algebra of finite dimension over a field P. An element r of A is said to be *nilpotent* if there exists a natural number k such that $r^k = 0$. The set R of all nilpotent elements of the algebra A is called the *radical* of the algebra A. If R consists of zero alone, A is called a *semisimple* algebra.

Proposition 3 *(Dedekind).* *A semisimple commutative associative algebra A of finite dimension over a field P is the direct sum*

$$A = A_1 \dotplus \cdots \dotplus A_t, \tag{1.11}$$

in which each term A_j is a field and a finite extension of the field P. Here, $A_i A_j = 0$ for $i \neq j$ ([4], Part 2).

Corollary. *The algebra A contains a unit element.*

Proposition 4 *(Wedderburn).* *If A is a finite-dimensional associative algebra over a perfect field P, then $A = R \dotplus B$*

is the direct sum of the radical R and the semisimple algebra B [12].[2]

Corollary. *The factor algebra A/R is isomorphic to B.*

Proposition 5 (*Frobenius*). *Let A denote a commutative associative algebra of arbitrary (not necessarily finite) dimension over a field P of characteristic zero.*[3] *If there exists a natural number v such that $a^v = 0$ for arbitrary $a \in A$, then $A^v = (0)$; that is, the product of v arbitrary elements of the algebra A is equal to 0.*

If $A^v = (0)$ but $A^{v-1} \neq (0)$, we shall say that the algebra A is *nilpotent of class v.*

Corollary. *If A is a nilpotent associative commutative algebra of class v, there exists an $a \in A$ such that $a^{v-1} \neq 0$.*

Lemma 2. *Let P denote an arbitrary field and let A denote a semisimple commutative subalgebra of P_n. Suppose that the unit element of the algebra A coincides with the unit element E_n of the algebra P_n. Then,*

(i) *A is a completely reducible set of operators;*

(ii) *the restriction of A to each irreducible invariant subspace Q_j, where $j = 1, \ldots, s$, is a field K_j whose degree over P coincides with the dimension of Q_j;*

(iii) *K_j is isomorphic to one of the terms A_i in (1.11), and, for every subalgebra A_i in (1.11), there exists a field K_j isomorphic to it.*

---

[2] In particular, P is an arbitrary number field.

[3] If the characteristic p of the field P is nonzero, then Proposition 5 is valid when $p > v$.

Proof: Since $E_n \in A$, we have $P^n = AP^n = (A_1 \dotplus \cdots \dotplus A_l)P^n$. Also, $A_i P^n = \sum_j A_i U_j$, where $u_1, \dots, u_n$ is a basis for P^n. The space $A_i u_j$ is an irreducible invariant subspace of a space P^n with respect to A. To see this, note that, for $a \in A$, we have $a = \sum_k a_k$, where $a_k \in A_k$ and consequently

$$a(A_i u_j) = (\sum_k a_k)(A_i u_j) = (a_i A_i)u_j \subset A_i u_j.$$

This means that $A_i u_j$ is invariant with respect to A. Furthermore, if v is a nonzero element of $A_i u_j$, then $Av = A_i v = A_i a_i u_j = A_i u_j$, where a_i is a nonzero element of A_i. Consequently, $A_i u_j$ is irreducible.

Thus, P^n is the sum of irreducible invariant subspaces Q_ν with respect to A of the form $A_i u_j$:

$$P^n = \sum_\nu Q_\nu. \tag{1.12}$$

Since the intersection of an irreducible invariant subspace $A_i u_j$ with an arbitrary invariant subspace with respect to A is equal to either (0) or $A_i u_j$, by removing the unnecessary terms from (1.12) we obtain the direct sum. This completes the proof of (i).

For $a = \sum_k a_k$, where $a_k \in A_k$, we have $a(b_i u_j) = a_i b_i u_j$ for $b_i \in A_i$. Consequently, the restriction of the algebra A to the subspace $A_i u_j$ is isomorphic to the field A_i. On the basis of Corollary 2 to Schur's lemma, the dimension of the subspace $A_i u_j$ coincides with the degree of the field A_i over P.

On the basis of Propositions 1 and 2, for every term A_i in (1.11) in the decomposition of P^n as a direct sum of the form (1.12), there exists a term Q_j for which the restriction of A is isomorphic to A_i. This completes the proof of the lemma.

We shall call a matrix g over a field P a *d-matrix* if g is similar to a diagonal matrix over some extension of the field P.

Theorem 1. *Let M denote the set of pairwise-commutative d-matrices in P_n. Then, over some finite extension of the field P, all the matrices in M can be simultaneously reduced to a diagonal form by a similarity transformation.*

Proof: Let us choose in M a maximal linearly independent system M_0 of matrices. Obviously, if the matrices M_0 in some extension Σ of the field P are reduced to diagonal form, then all the matrices M are also reduced to diagonal form. Therefore, it will be sufficient to carry out the proof for a finite set M_0.

If Σ is the splitting field of the minimum polynomials of the matrices of the set M_0, then, over Σ, every matrix in M_0 is similar to some diagonal matrix. Let us now show that over the field Σ, it is possible to simultaneously transform all matrices in M_0 to diagonal form. We use induction on n. Obviously, the case which requires proof is that in which M_0 has a nonscalar matrix a. Let us use a similarity transformation over the field Σ to reduce a to the form $a = [\lambda_1 E_{n_1}, \ldots, \lambda_s E_{n_s}]$ for $\lambda_i \in \Sigma$, where $\lambda_i \neq \lambda_j$ for $i \neq j$. If $b \in M$, the condition $ab = ba$ implies that $b = [B_1, \ldots, B_s]$, where $B_i \in \sum n_i$ (cf. Section 1, Corollary 6). Since b is a d-matrix, the minimum polynomial $\varphi(x)$ of the matrix b has no repeated roots. But the minimum polynomials of the matrices B_i are divisors of the polynomial $\varphi(x)$. Therefore, they too have no repeated roots and, consequently, the B_i are d-matrices. The theorem then follows by induction.

It follows from Theorem 1, in particular, that the product of d-matrices belonging to an Abelian group is a d-matrix. More generally, if the base field is perfect, then the product of d-matrices belonging to a locally nilpotent group is a d-matrix [*31*]. However, this theorem is not valid for solvable groups [*32*].

The antipode of a d-matrix is a unipotent matrix, that is, a matrix all of whose eigenvalues are equal to unity. In a solvable linear group over a field of characteristic zero, the set of all unipotent matrices is a normal subgroup. In particular, the product of unipotent matrices of a solvable group over a field of zero characteristic is a unipotent matrix [*33*].

Corollary. *If the field P is perfect and A is a semisimple commutative subalgebra of a complete linear algebra P_n, then, over some finite extension of the field P, all the matrices in A can be simultaneously reduced to diagonal form by a similarity transformation.*

Proof: It will be sufficient to show that A consists of d-matrices. Let us represent A in the form (1.11). Then, for arbitrary $a \in A$, we have

$$a = \sum_i a_i,$$

where a_i is an element of the field A_i. The minimum polynomial of an element of the field is irreducible and, since the field P is perfect, the minimum polynomial $\varphi_i(x)$ of an element a_i does not have repeated roots. The minimum polynomial of the element a is the least common multiple of the polynomials $\varphi_i(x)$ and, consequently, it too has no repeated roots. Therefore, a is a d-matrix.

Converse.   *The commutative ring of d-matrices over an arbitrary field is semisimple.*

Obviously, the only nilpotent d-matrix is the zero matrix.

In the case of an arbitrary set of commutative matrices, instead of Theorem 1 we have

Theorem 2.   *Let P denote an arbitrary field and let M denote the set of pairwise-commutative matrices in P_n. There exist a finite extension Σ of the field P and a similarity transformation over Σ that reduces all matrices in M simultaneously to triangular form. In particular, if all roots of the characteristic polynomial of each matrix in M belong to the field P, we may take P as the field Σ.*

Proof: Let M_0 denote a maximal subset of matrices in M that are linearly independent over P. It will be sufficient to prove the theorem for a finite set M_0. Suppose now that Σ is the splitting field of the minimum polynomials of all the matrices in M_0. Let us show that all the matrices in M_0 are simultaneously reduced to triangular form over Σ. We shall prove this by induction on n. If $n = 1$, the conclusion of the theorem is obvious. Suppose that $n > 1$ and that the conclusion is valid for dimension $m < n$. If all the matrices in M_0 are scalar matrices, the assertion is obvious. Suppose that M_0 contains a nonscalar matrix a and suppose that $\lambda \in \Sigma$ is an eigenvalue of it. Consider the subspace $Q \subset \Sigma^n$ consisting of all vectors q such that $a(q) = \lambda q$. If $b \in M_0$, then $bQ \subset Q$. This is true because $ab(q) = ba(q)$ for $q \in Q$. Therefore, $a(b(q)) = \lambda b(q)$ for $b(q) \in Q$. Consequently, Q is an invariant subspace of the space Σ^n with respect to M_0. Obviously, $(0) \neq Q \neq \Sigma^n$. Let $v_1, \ldots, v_t$ denote a basis of Q and let

$$u_1, \ldots, u_s, v_1, \ldots, v_t \qquad (1.13)$$

denote a basis of Σ^n. Then, in the basis (1.13), the matrices M_0 assume the form

$$m = \begin{bmatrix} a(m) & 0 \\ c(m) & b(m) \end{bmatrix},$$

where $a(m)$ is of degree s and $b(m)$ is of degree t for $s, t < n$. From this and the induction hypothesis, it follows that M_0 can be reduced to triangular form over Σ. This completes the proof of the theorem.

Corollary 7. *Let P denote an algebraically closed field and let M denote the set of pairwise-commutative matrices in P_n. Then, all matrices in M can be simultaneously reduced to triangular form by a similarity transformation over P.*

Corollary 8 (Frobenius). *Let P denote an arbitrary field and let $a_1, \ldots, a_k$ denote pairwise-commutative matrices in P_n. The eigenvalues of the matrices a_i can be indexed $\lambda_{i1}, \ldots, \lambda_{in}$, for $i = 1, \ldots, k$, in such a way that, for an arbitrary polynomial $f(x_1, \ldots, x_k)$ over P, the eigenvalues of the matrix $f(a_1, \ldots, a_k)$ are the numbers $f(\lambda_{1j}, \ldots, \lambda_{kj})$ for $j = 1, \ldots, n$.*

Proof: Over some field $\Sigma \supset P$, the matrices a_i, for $i = 1, \ldots, k$, can be reduced to a common triangular form with numbers λ_{ij}, for $j = 1, \ldots, k$, on the principal diagonal. The matrix $f(a_1, \ldots, a_k)$ in the same basis is also triangular, and its diagonal consists of the numbers $f(\lambda_{1i}, \ldots, \lambda_{ki})$.

We note that Corollary 2 is valid under arbitrary conditions when the matrices a_i, for $i = 1, \ldots, k$, have a common triangular form. For two matrices, the same statement can be given a more exact formulation:

The eigenvalues of two matrices a and b belonging to P_n can be numbered α_i, β_j in such a way that, for an arbitrary polynomial $f(x, y)$ over P, the eigenvalues of the matrix $f(a, b)$ are the numbers $f(\alpha_i, \beta_i)$ if and only if the matrices a and b have a common triangular form [33].

Theorems 3 and 4 supplement Theorem 2, giving more precise information.

Theorem 3. *Let P denote an arbitrary field and let M denote the set of pairwise-commutative operators in P_n. Then, the space P^n can be represented as the direct sum of invariant subspaces Q_j with respect to M for $j = 1, \ldots, k$, such that the irreducible parts of the restriction $M|Q_j$ are equivalent and, for $j \neq i$, the irreducible parts $M|Q_i$ and $M|Q_j$ are not equivalent.*

To prove this, we need

Lemma 3. *Let M denote a set consisting of commutative matrices m of the form*

$$m = \begin{bmatrix} a(m) & 0 \\ b(m) & c(m) \end{bmatrix},$$

where $m \to a(m)$ and $m \to c(m)$ are irreducible nonequivalent representations of the set M, whose degrees are, respectively, μ and v. Then, there exists a matrix

$$t = \begin{bmatrix} E_\mu & 0 \\ s & E_v \end{bmatrix},$$

such that $t^{-1}mt = [a(m), c(m)]$ for all $m \in M$.

Proof: Let A denote the algebra of matrices that is generated by the set M.

The matrices g of the algebra A are of the form

$$g = \begin{bmatrix} a(g) & 0 \\ b(g) & c(g) \end{bmatrix}, \tag{1.14}$$

where $a(g)$ and $c(g)$ range over the commutative irreducible algebras A_1 and C_1, respectively, where A_1 and C_1 are fields (cf. Section 1, Corollary 2). Consider the radical R of the algebra A. Since $a(g)$ and $c(g)$ in (1.14) are elements of fields, an arbitrary nilpotent element of the algebra A is of the form

$$r = \begin{bmatrix} 0 & 0 \\ b & 0 \end{bmatrix}.$$

It follows from the equation $bg = gb$, where g is an arbitrary matrix of the form (1.14), that $ba(g) = c(g)b$ for $g \in A$. From Schur's Lemma, $b = 0$, $r = 0$, and A is semisimple, consequently, completely reducible (cf. Lemma 2). Let

$$u_1, \ldots, u_\mu, v_1, \ldots, v_\nu \tag{1.15}$$

denote that basis for the space P^n in which the matrices of the algebra A are of the form of Eq. (1.14).

Then, the subspace $Q_2 = [v_1, \ldots, v_\nu]$ is invariant with respect to A. By virtue of Proposition 1, for Q_2 there exists an invariant complement $Q_1 = [w_1, \ldots, w_\mu]$. In the basis

$$w_1, \ldots, w_\mu, v_1, \ldots, v_\nu \tag{1.16}$$

the matrices $g \in A$ are of the form $g = [d^{-1}a(g)d, c(g)]$, where

$$\begin{bmatrix} d & 0 \\ b & E_\nu \end{bmatrix} = f$$

is the matrix of the transformation from (1.15) to (1.16). Now, if $c = [d^{-1}, E_\nu]$ and $t = cf$, then

$$t = \begin{bmatrix} E_\mu & 0 \\ s & E_\nu \end{bmatrix} \qquad \text{and} \qquad t^{-1}gt = [a(g), c(g)]$$

for $g \in A$. This completes the proof of the lemma.

Proof of Theorem 3: Obviously, the matrices of the set M can be simultaneously converted to the form

$$m = \begin{bmatrix} a_1(m) & 0 & \ldots & 0 \\ a_{21}(m) & a_2(m) & \ldots & 0 \\ \ldots\ldots\ldots\ldots\ldots\ldots\ldots \\ a_{s1}(m) & a_{s2}(m) & \ldots & a_s(m) \end{bmatrix}, \qquad (1.17)$$

where $m \to a_i(m)$ is an irreducible representation M_i of degree n_i for $i = 1, \ldots, s$, where $\sum n_i = n$. If all the representations M_i are equivalent, the theorem is trivial. On the other hand, if not all the M_i are equivalent, there exist "adjacent" nonequivalent representations M_j and M_{j+1}. Then, consider the representation

$$m \to m_j = \begin{bmatrix} a_j(m) & 0 \\ a_{jj+1}(m) & a_{j+1}(m) \end{bmatrix}.$$

Since the set of all the matrices m_j is commutative, it follows from Lemma 3 that there exists a matrix

$$t_j = \begin{bmatrix} E_{n_j} & 0 \\ s_j & E_{n_{j+1}} \end{bmatrix},$$

transforming all the m_j to the form

$$m_j = [a_j(m), a_{j+1}(m)]. \tag{1.18}$$

If $t = [E_{n_1 + \cdots + n_{j-1}}, t_j, E_{n_{j+2} + \cdots + n_s}]$, then, after the transformation of the matrices (1.17) by t, the $a_{jj+1}(m)$ will be replaced by zeros for all $m \in M$ but the $a_{kk+1}(m)$ will remain unchanged for $k \neq j$. Therefore, we may assume that $a_{jj+1}(m) = 0$ in (1.17) if M_j and M_{j+1} are nonequivalent.

Now, if $a_{ij}(m) = 0$ for all $m \in M$ when M_i and M_j are nonequivalent and $j < k + i$ (where $k > 1$), let us consider the representation

$$m \to m_{ij} = \begin{bmatrix} a_i(m) & 0 \\ a_{ij}(m) & a_j(m) \end{bmatrix}$$

where $j = i + k$. It is easy to see that the matrices m_{ij} constitute a commutative set. Therefore, according to Lemma 3, there exists a matrix

$$t_{ij} = \begin{bmatrix} E_{n_i} & 0 \\ s_{ij} & E_{n_j} \end{bmatrix},$$

transforming all the matrices m_{ij} to the form (1.18). If

$$t = [E_{n_1 + \cdots + n_{i-1}}, T_{ij}, E_{n_{j+1} + \cdots + n_s}],$$

where

$$T_{ij} = \begin{bmatrix} E_{n_i} & & \\ 0 & E_{n_{i+1}} & \\ \cdots\cdots\cdots\cdots\cdots & \\ s_{ij} & 0 & E_{n_j} \end{bmatrix},$$

then, after the transformation of the matrices (1.17) by t, the $a_{ij}(m)$ will be replaced by zeros and the other $a_{pl}(m)$, where $l \leqslant p + k$, will remain unchanged. Consequently, we may assume that $a_{ij}(m) = 0$ in (1.17) for all $m \in M$ whenever M_i and M_j are not equivalent. But then, by applying a suitable substitution to the basis P^n, we can put the equivalent parts of M in a sequence. This completes the proof of the theorem.

Corollary. Let P denote an arbitrary field and let M denote the set of pairwise-commutative operators in P_n. Then, P^n can be represented as a direct sum of subspaces Q_j (for $j = 1, \ldots, s$) that are invariant with respect to M such that, for arbitrary $f \in M$, the minimum polynomial of the restriction $f|Q_j$ is a power of a polynomial that is irreducible over P.

To see this, note that the same subspaces Q_j can be chosen as in the preceding theorem.

Theorem 4. Let P denote a perfect field and let M denote a set of commutative matrices. Suppose that the irreducible parts of M are all equivalent. Then, all the matrices of M can be simultaneously reduced by a similarity transformation to the form

$$
m = \begin{bmatrix}
a(m) & 0 & \ldots & 0 \\
a_{21}(m) & a(m) & \ldots & 0 \\
\multicolumn{4}{c}{\dotfill} \\
a_{s1}(m) & a_{s2}(m) & \ldots & a(m)
\end{bmatrix}, \tag{1.19}
$$

where $m \to a(m)$ is an irreducible representation of the set M of degree $r = ns^{-1}$ and the matrices $a(m)$ and $a_{\alpha\beta}(m)$ are elements of the field Ω of degree r over P.

Proof: All the matrices $m \in M$ can be written simultaneously in the form

$$m = \begin{bmatrix} b(m) & 0 & \dots & 0 \\ a_{21}(m) & b(m) & \dots & 0 \\ \dots\dots\dots\dots\dots\dots\dots\dots\dots \\ a_{s1}(m) & a_{s2}(m) & \dots & b(m) \end{bmatrix}, \qquad (1.20)$$

where $m \to b(m)$ is an irreducible representation of the set M by means of matrices in P_r for $r = ns^{-1}$. Together with M, let us look at the algebra A generated by the matrices M. By virtue of Proposition 4, the algebra A is a direct sum

$$A = R \dotplus B, \qquad (1.21)$$

where R is a radical and B is a semisimilar algebra isomorphic to A/R. The matrices of the algebra A are all of the form (1.20), where $b(m)$ takes on values in some field Σ of degree r over P as m ranges over A (cf. Section 1, Corollary 2). A field contains no divisors of zero. Consequently, R consists of all matrices of the algebra A that are of the form (1.20) with $b(m) = 0$.

Let us consider a homomorphism γ of the algebra A onto the algebra $D \subset P_n$ of all matrices $[b(m),\dots,b(m)]$, where $b(m) \in \Sigma$: $\gamma(m) = [b(m),\dots,b(m)]$. Obviously, R is the kernel of the homomorphism γ. Consequently, $A/R \cong D \cong \Sigma$. Thus, $B \cong \Sigma$. On the basis of Lemma 1 and Corollary 2 of Section 1, B is a completely reducible set of matrices, all irreducible parts of B are equivalent, and every irreducible part of B is a field of degree r over P. Let us now reduce all matrices $c \in B$ to the form $c = [a(c),\dots,a(c)]$, where the matrix $a(c)$ ranges over a field Q of degree r over P.

Let us return to the radical R. Every matrix in it commutes with every matrix in B. Therefore, all matrices $r \in R$ are of the form

$$\rho = \begin{bmatrix} b_{11}(\rho) & \cdots & b_{1s}(\rho) \\ \cdots\cdots\cdots\cdots\cdots \\ b_{s1}(\rho) & \cdots & b_{ss}(\rho) \end{bmatrix},$$

where $b_{\alpha\beta}(\rho)$ belongs to the centralizer Z of the field Q in P_r. Let $Z = Q$ (cf. Section 1, Corollary 2). Therefore, $b_{\alpha\beta}(\rho) \in Q$.

For the moment, let us write the matrices B and R as elements of the full linear algebra Q_s. R is a commutative algebra of matrices, the roots of whose characteristic polynomials are all zero. This means that all the matrices of R can be simultaneously reduced by a similarity transformation over the field Q to triangular form

$$\rho = \begin{bmatrix} 0 & 0 & \cdots & 0 \\ a_{21}(\rho) & 0 & \cdots & 0 \\ \cdots\cdots\cdots\cdots\cdots\cdots \\ a_{s1}(\rho) & a_{s2}(\rho) & \cdots & 0 \end{bmatrix}, \qquad a_{\alpha\beta}(\rho) \in Q.$$

The matrices B are scalar matrices and therefore are not altered by the similarity transformation. On the basis of (1.21), all matrices of the algebra A are now of the form

$$m = \begin{bmatrix} a(m) & 0 & \cdots & 0 \\ a_{21}(m) & a(m) & \cdots & 0 \\ \cdots\cdots\cdots\cdots\cdots\cdots \\ a_{s1}(m) & a_{s2}(m) & \cdots & a(m) \end{bmatrix},$$

where $a(m)$ and $a_{\alpha\beta}(m)$ belong to Q. If we now return to the original representation of the elements of the field Q with the aid of the matrices P_r, we obtain the form (1.19). This completes the proof of the theorem.

3. Matrices That Commute with a Given Matrix

For a given field P, we shall denote by $P[x]$ the ring of all polynomials $f(x)$ over the field P. Similarly, for a square matrix a over P, we denote by $P[a]$ the ring of all matrices that can be represented in the form $f(a)$, where $f(x) \in P[x]$.

Theorem 5. *Let a denote a member of P_n. The centralizer of a in P_n coincides with the ring $P[a]$ if and only if the minimum polynomial of a coincides with the characteristic polynomial of a.*

Proof: Let us choose a basis for P^n such that the matrix of the operator a assumes the general normal form

$$a = [c_1, \ldots, c_k], \tag{1.22}$$

where c_j is the matrix accompanying the polynomial $\varphi_j(x)$ and $\varphi_1(x), \ldots, \varphi_k(x)$ are the invariant factors of a. As we know, $\varphi_k(x)$ is the minimum polynomial of a.

Suppose that $\varphi_k(x)$ is different from the characteristic polynomial of a. Then, $k > 1$ in (1.22). The matrix $b = [E, 0, \ldots, 0]$, all of whose diagonal blocks are of the same dimensions as the corresponding blocks of a, commutes with a. However, we cannot represent b in the form $b = f(a)$, where $f(x) \in P[x]$. Indeed, if $b = f(a)$, then

$$f(c_1) = E, \tag{1.23}$$

$$f(c_k) = 0. \tag{1.24}$$

It follows from (1.24) that $\varphi_k(x)$ divides $f(x)$ since $\varphi_k(x)$ is the minimum polynomial of c_j for $j = 1, \ldots, k$. But $\varphi_1(x)$ divides $\varphi_k(x)$. Consequently, $f(c_1) = 0$. This last equation contradicts (1.23). This completes the proof of the necessity.

Now, suppose that the minimum and characteristic polynomials of a coincide. Then, $k = 1$ in (1.22) and

$$a = c_1 = \begin{bmatrix} 0 & 0 & \ldots & 0 & -\alpha_0 \\ 1 & 0 & \ldots & 0 & -\alpha_1 \\ 0 & 1 & \ldots & 0 & -\alpha_2 \\ \multicolumn{5}{c}{\cdots\cdots\cdots\cdots\cdots\cdots\cdots} \\ 0 & 0 & \ldots & 1 & -\alpha_{n-1} \end{bmatrix} \qquad (1.25)$$

where $\alpha_0 + \alpha_1 x + \cdots + \alpha_{n-1} x^{n-1} + x^n = \varphi_1(x)$. Let $u_0, u_1, \ldots,$ u_{n-1} denote that basis for P^n in which the matrix a is of the form (1.25). Then, $au_0 = u_1, au_1 = u_2, \ldots, au_{n-2} = u_{n-1}$. For $b \in P_n$, we obviously have

$$bu_0 = \lambda_0 u_0 + \lambda_1 u_1 + \cdots + \lambda_{n-1} u_{n-1}$$
$$= \lambda_0 E_n + \lambda_1 a + \cdots + \lambda_{n-1} a^{n-1} u_0 = f(a) u_0,$$
$$f(x) = \lambda_0 + \lambda_1 x + \cdots + \lambda_{n-1} x^{n-1} \in P[x].$$

Now, if $ba = ab$, then $b_1 = b - f(a)$ also commutes with a and we have $b_1 u_0 = 0$ and $b_1 u_s = 0$ for $s = 1, \ldots, n - 1$. Consequently, $b_1 = 0$ and $b = f(a)$. This completes the proof of the theorem.

Now, let a denote an arbitrary operator in P_n. Consider the algebra $ZP_n(a)$, the centralizer of a in P_n. We have already seen that, if the minimum and characteristic polynomials of a coincide, then $ZP_n(a) = P[a]$. Let us denote by $\varphi(x)$ the minimum polynomial of the operator a and let $\varphi(x) = q_1(x) \cdots q_k(x)$, where $q_j(x) = p_j^{\alpha_j}(x)$ is the decomposition of $\varphi(x)$ as the product of powers of distinct polynomials $p_j(x)$ that are irreducible in $P[x]$. Then, as is well known from linear algebra, we have

Proposition 6. *The space P^n in which a operates can be represented as a direct sum*

$$P^n = Q_i \dotplus \cdots \dotplus Q_k \qquad (1.26)$$

of invariant subspaces $Q_1, \ldots, Q_n$ with respect to a such that the minimum polynomial of the restriction $a|Q_j$ is equal to $q_j(x) = p_j^{\alpha_j}(x)$. The direct summand Q_j of the decomposition (1.26) is the set of all vectors $v \in P^n$ such that $q_j(a)v = 0$.

Lemma 4. (i) *Every subspace Q_j in the decomposition (1.26) is invariant with respect to $ZP_n(a)$.*

(ii) *The algebra $ZP_n(a)$ is the direct sum of k rings $ZP_{n_j}(a_j)$, where n_j is the dimension of Q_j and $a_j = a|Q_j$, for $j = 1, \ldots, k$. Here $n_1 + \cdots + n_k = n$.*

Proof: Let b denote an element of $ZP_n(a)$ and let v denote an element of Q_j. Then, $ab(v) = ba(v) = q_j(a)b(v) = bq_j(a)(v) = 0$. Consequently, $b(v) \in Q_j$, where Q_j is invariant with respect to $ZP_n(a)$. This proves (i).

It follows from (i) that, in a basis of P^n in which we obtain decomposition (1.26), the matrices b in $ZP_n(a)$ are of the form

$$b = [b_1, \ldots, b_k], \qquad (1.27)$$

where

$$b_j \in ZP_{n_j}(a_j).$$

On the other hand, an arbitrary matrix of the form (1.27) belongs to $ZP_n(a)$. Consequently, (ii) holds.

Thus, the study of $ZP_n(a)$ has been reduced to the case in which the minimum polynomial a is a power of a polynomial that is irreducible in $P[x]$.

We present one more simple lemma regarding the extension of the basic field. If Ω is an extension of the field P, then $a \in P_n$ implies $a \in \Omega_n$. Consequently, we may speak of the *centralizer* $Z\Omega_n(a)$ of the operator a in Ω_n.

Lemma 5. *The dimension of the algebra $Z\Omega_n(a)$ over Ω is equal to the dimension of $ZP_n(a)$ over P. A basis $ZP_n(a)$ over P is a basis for $Z\Omega_n(a)$ for Ω.*

Proof: $ZP_n(a)$ is the set of solutions of the equation

$$ax = xa, \tag{1.28}$$

where $x \in P_n$. Obviously, Eq. (1.28) is equivalent to some system F of homogeneous linear equations with n^2 unknowns x_{ij}, where $x = \|x_{ij}\|$. Thus, the dimension of $ZP_n(a)$ over P is determined by the rank of the system F. Obviously, the rank of the system F remains unchanged when we shift from P to Ω. The conclusion of the lemma then follows.

Let us now find the form of the matrices of the ring $ZP_n(a)$ for the case in which all the roots of the characteristic polynomial a belong to the field P. As a preliminary, we give a definition:

A $\mu \times \nu$ matrix c over a field P is said to be *triangularly striped* if

(i)   for $\mu = \nu$,

$$c = \begin{bmatrix} \gamma_1 & 0 & \cdots & 0 \\ \gamma_2 & \gamma_1 & \cdots & 0 \\ \gamma_3 & \gamma_2 & \cdots & 0 \\ \cdots\cdots\cdots\cdots\cdots\cdots \\ \gamma_\mu & \gamma_{\mu-1} & \cdots & \gamma_1 \end{bmatrix},$$

(ii)   for $\mu < \nu$,

$$c = \begin{bmatrix} \gamma_1 & 0 & \ldots & 0 & 0 & \ldots & 0 \\ \gamma_2 & \gamma_1 & \ldots & 0 & 0 & \ldots & 0 \\ \gamma_3 & \gamma_2 & \ldots & 0 & 0 & \ldots & 0 \\ \cdots\cdots\cdots\cdots\cdots\cdots\cdots\cdots\cdots \\ \gamma_\mu & \gamma_{\mu-1} & \ldots & \gamma_1 & 0 & \ldots & 0 \end{bmatrix},$$

(iii)   for $\mu > \nu$,

$$c = \begin{bmatrix} 0 & 0 & \ldots & 0 \\ \vdots & \vdots & \cdots\cdots\cdots \\ 0 & 0 & \ldots & 0 \\ \gamma_1 & 0 & \ldots & 0 \\ \gamma_2 & \gamma_1 & \ldots & 0 \\ \gamma_3 & \gamma_2 & \ldots & 0 \\ \cdots\cdots\cdots\cdots\cdots\cdots \\ \gamma_\nu & \gamma_{\nu-1} & \ldots & \gamma_1 \end{bmatrix}, \quad \gamma_i \in P.$$

Obviously, if the elements of the first column of a triangularly striped matrix c are all zeros, then c is the zero matrix.

One can easily see that the space of all triangularly striped $\mu \times \nu$ matrices is of dimension min $\{\mu, \nu\}$.

We define

$$I_m(\lambda) = \begin{bmatrix} \lambda & 0 & \ldots & . & . & 0 \\ 1 & \lambda & \ldots & . & . & 0 \\ \cdots\cdots\cdots\cdots\cdots\cdots\cdots \\ 0 & 0 & \ldots & 0 & 1 & \lambda \end{bmatrix},$$

where $\lambda \in P$ and m is the degree of $I_m(\lambda)$.

Theorem 6.   *Suppose that*

$$a = [I_{m_1}(\lambda), \ldots, I_{m_t}(\lambda)],$$

where $m_1 + \cdots + m_t = n$.   *Then,* $ZP_n(a)$ *consists of all matrices* b *of the form*

$$b = \begin{bmatrix} b_{11} & b_{12} & \ldots & b_{1t} \\ b_{21} & b_{22} & \ldots & b_{2t} \\ \ldots\ldots\ldots\ldots\ldots\ldots \\ b_{t1} & b_{t2} & \ldots & b_{tt} \end{bmatrix},$$

where b_{ij} *is an arbitrary triangularly striped* $m_i \times m_j$ *matrix over* P.

Proof: The equation $ab = ba$ is equivalent to $a_0 b = b a_0$, where

$$a_0 = [I_{m_1}(0), \ldots, I_{m_t}(0)].$$

If

$$b = \begin{bmatrix} b_{11} & b_{12} & \ldots & b_{1t} \\ b_{21} & b_{22} & \ldots & b_{2t} \\ \ldots\ldots\ldots\ldots\ldots\ldots \\ b_{t1} & b_{t2} & \ldots & b_{tt} \end{bmatrix},$$

where b_{ij} is an $m_i \times m_j$ matrix, then commutativity of a_0 and b is equivalent to the conditions

$$b_{ij} I_{m_j} = I_{m_i} b_{ij}, \tag{1.29}$$

where $I_{m_\alpha} = I_{m_\alpha}(0)$ for $\alpha = j, i$.

We shall now show that Eq. (1.29) holds if and only if b_{ij} is a triangularly striped matrix. If $m_i = m_j$, then, on the basis of Theorem 5, b_{ij} satisfies condition (1.29) if and only if $b_{ij} = f(I_{m_i})$ for $f(x) \in P$. On the other hand, simple calculations show that $P[I_{m_i}]$ is the set of all triangularly striped matrices of degree m_i. Consequently, our assertion holds for $m_i = m_j$. Let us now use induction on the sum $m_i + m_j$. Suppose that $m_i < m_j$. Let us rewrite (1.29) in the form

$$[CD] \left[\begin{array}{c|c} I_{m_i} & 0 \\ \hline 0 \ 0 \ \ldots \ 0 \ 1 & \\ \hline 0 & I_{m_j - m_i} \end{array} \right] = I_{m_i}[CD], \qquad (1.30)$$

where C consists of the first m_i columns of the matrix b_{ij} and D consists of the remaining columns. Equation (1.30) is equivalent to the system

$$[CD] \left[\begin{array}{c} I_{m_i} \\ \hline 0 \ 0 \ \ldots \ 1 \\ \hline 0 \end{array} \right] = I_{m_i} C, \qquad (1.31)$$

$$DI_{m_j - m_i} = I_{m_i} D. \qquad (1.32)$$

By induction, all triangularly striped matrices and only these satisfy condition (1.32). Let us rewrite condition (1.31) in the form

$$[CD] \left(\left[\begin{array}{c} I_{m_i} \\ \\ 0 \end{array} \right] + \left[\begin{array}{c} 0 \\ \hline 0 \ 0 \ \ldots \ 1 \\ \hline 0 \end{array} \right] \right)$$

$$= CI_{m_i} + D \left[\begin{array}{c} 0 \ 0 \ \ldots \ 1 \\ \hline 0 \end{array} \right] = I_{m_i} C. \qquad (1.33)$$

Suppose that $b_{ij} = ||\beta_{\alpha\beta}||$. Then, (1.33) can be written

$$
\begin{bmatrix}
\beta_{12} & \beta_{13} & \cdots & \beta_{1m_i} & 0 \\
\beta_{22} & \beta_{23} & \cdots & \beta_{2m_i} & 0 \\
\multicolumn{5}{c}{\dotfill} \\
\beta_{m_i2} & \beta_{m_i3} & \cdots & \beta_{m_im_i} & 0
\end{bmatrix}
+
\begin{bmatrix}
0 & \cdots & 0 & \beta_{1m_{i+1}} \\
0 & \cdots & 0 & \beta_{2m_{i+1}} \\
\multicolumn{4}{c}{\dotfill} \\
0 & \cdots & 0 & \beta_{m_im_{i+1}}
\end{bmatrix}
$$

$$
=
\begin{bmatrix}
0 & 0 & \cdots & 0 \\
\beta_{11} & \beta_{12} & \cdots & \beta_{1m_i} \\
\multicolumn{4}{c}{\dotfill} \\
\beta_{m_{i-1}1} & \cdots & \cdots & \beta_{m_{i-1}m_i}
\end{bmatrix}.
\tag{1.34}
$$

Now, we are interested only in the first column of the matrix D, that is, in the elements $\beta_{\nu m_{i+1}}$ for $\nu = 1, \ldots, m_i$. From (1.34), we first obtain $0 + \beta_{1m_{i+1}} = 0$ and, consequently, $\beta_{1m_{i+1}} = 0$. Also from (1.34), we find $\beta_{2m_{i+1}} = \beta_{1m_i}$. But $\beta_{1m_i} = 0$. Consequently, $\beta_{2m_{i+1}} = 0$, $\beta_{3m_{i+1}} = \beta_{2m_i} = \beta_{1m_{i-1}} = 0$, $\beta_{4m_{i+1}} = \beta_{3m_i} = \beta_{2m_{i-1}} = \beta_{1m_{i-2}} = 0$. Proceeding in this way, we see that all the $b_{\nu\mu_{i+1}} = 0$. Thus, the first column of the matrix D consists only of zeros. Since here D is triangularly striped, it follows that D is the zero matrix. Condition (1.33) now takes the form

$$
CI_{m_i} = I_{m_i}C.
\tag{1.35}
$$

All triangularly striped matrices C and only these satisfy condition (1.35). Consequently, $b_{ij} = [C0]$ is an arbitrary triangularly striped matrix.

For $m_i > m_j$, the proof is analogous. This completes the proof of the theorem.

Corollary 9. *If* $a = [I_{m_1}(\lambda), \ldots, I_{m_t}(\lambda)]$ *and* $m_1 \geqslant m_2 \geqslant \cdots \geqslant m_t$, *then the dimension of the algebra* $ZP_n(a)$ *over* P *is equal to the number* $m_1 + 3m_2 + \cdots + 2(t-1)m_t$.

Proof: The dimension of the space of triangularly striped matrices b_{11} (cf. Theorem 6) is equal to m_1, the dimension of the space of all matrices of the form

$$\begin{bmatrix} 0 & b_{12} \\ b_{21} & b_{22} \end{bmatrix}$$

is equal to $3m_2$, etc. The assertion of the corollary then follows.

Corollary 10. *For arbitrary* $a \in P_n$, *the dimension of* $ZP_n(a)$ *over* P *is not less than* n. *The dimension of* $ZP_n(a)$ *over* P *is equal to* n *if and only if the minimum and characteristic polynomials of* a *coincide.*

This follows from the preceding corollary and Lemmas 4 and 5.

Lemma 6. *Suppose that* $a = [I_{m_1}(\lambda), \ldots, I_{m_t}(\lambda)]$. *Then, the center* Z *of the algebra* $ZP_n(a)$ *coincides with the algebra* $P[a]$.

Proof: Obviously, we may assume that $m_1 \geqslant m_2 \geqslant \cdots \geqslant m_t$. In the algebra $ZP_n(a)$ there are matrices π_j of the form $\pi_j = [0_{m_1}, \ldots, E_{m_j}, \ldots, 0_{m_t}]$, for $j = 1, \ldots, t$. If $z \in Z$, then $z\pi_j = \pi_j z$. Therefore, $z = [z_1, \ldots, z_j, \ldots, z_t]$, where z_j is an $m_j \times m_j$ matrix over P for $j = 1, \ldots, t$. Since $za = az$, we have $z_j I_{m_j}(\lambda) = I_{m_j}(\lambda) z_j$. Consequently, z_j is a triangularly striped $m_j \times m_j$ matrix. Let us show that the block z_1 completely determines the matrix z since the block z_j is obtained from z_1 by removing the rows and columns of index greater than m_j. Suppose that

$$b = \begin{bmatrix} 0 & 0 & \ldots & 0 \\ \cdots\cdots\cdots\cdots\cdots \\ 0 & 0 & \ldots & 0 \\ b_{j1} & 0 & \ldots & 0 \\ \cdots\cdots\cdots\cdots\cdots \\ 0 & 0 & \ldots & 0 \end{bmatrix},$$

where $b_{j1} = [E_{m_j} 0]$ is an $m_j \times m_1$ matrix.   Obviously, $b \in ZP_n(a)$.   Consequently, $bz = zb$.   Therefore, $b_{j1}z_1 = z_j b_{j1}$ or

$$[E_{m_j} 0]z_1 = z_j [E_{m_j} 0]. \tag{1.36}$$

Setting

$$z_1 = \begin{bmatrix} \gamma_1 & 0 & \ldots & 0 \\ \gamma_2 & \gamma_1 & \cdots & 0 \\ \cdots\cdots\cdots\cdots\cdots\cdots \\ \gamma_{m_1} & \gamma_{m-1} & \cdots & \gamma_1 \end{bmatrix},$$

we obtain from (1.36)

$$z_j[E_{m_j} 0] = [z_j \, 0] = [E_{m_j} 0]z_1 = \left[\begin{bmatrix} \gamma_1 & 0 & \cdots & \\ \gamma_2 & \gamma_1 & \cdots & \\ \cdots\cdots\cdots\cdots\cdots & \\ \gamma_{m_j} & \cdots & \cdots & \gamma_1 \end{bmatrix} 0 \right].$$

Consequently,

$$z_j = \begin{bmatrix} \gamma_1 & 0 & \ldots & 0 \\ \gamma_2 & \gamma_1 & \cdots & 0 \\ \cdots\cdots\cdots\cdots\cdots \\ \gamma_{m_j} & \cdots & \cdots & \gamma_1 \end{bmatrix}.$$

Thus, z_1 completely determines z. From this it follows that the dimension of Z over Γ is equal to m_1. But the dimension of $P[a]$ is also equal to m_1 since m_1 is the degree of the minimum polynomial of a. This completes the proof of the lemma since $P[a] \subset Z$.

Lemma 7. *If $a \in P_n$ can be reduced to a Jordan normal form over P, then the center Z of the algebra $ZP_n(a)$ coincides with $P[a]$.*

Proof: Suppose that the condition of the lemma is satisfied. Then, the minimum polynomial $\varphi(x)$ of the matrix a is of the form $\varphi(x) = (x - \lambda_1)^{\alpha_1} \cdots (x - \lambda_k)^{\alpha_k}$, where $\lambda_j \in P$ and $\lambda_i \neq \lambda_j$ for $i \neq j$. In accordance with Lemma 4, there exists a decomposition $P^n = Q_1 \dotplus \cdots \dotplus Q_k$ such that the minimum polynomial of the restriction $a|Q_j$ is equal to $(x - \lambda_j)^{\alpha_j}$, Q_j is invariant with respect to $ZP_n(a)$, and $ZP_n(a)$ is the direct product of its restrictions $ZP_n(a)|Q_j$. Obviously, the center Z of the algebra $ZP_n(a)$ is the direct product of the centers Z_j of the algebras $ZP_n(a)|Q_j$ for $j = 1, \ldots, k$. Consequently, in accordance with Lemma 6, the dimension of Z over P is equal to $\alpha_1 + \alpha_2 + \cdots + \alpha_n$; that is, it coincides with the degree of the minimum polynomial $\varphi(x)$ of the matrix a. The conclusion of the lemma follows from this.

Theorem 7. *Let P denote an arbitrary field and let a denote an arbitrary matrix in P_n. Then, the center Z of the algebra $ZP_n(a)$ coincides with the algebra $P[a]$.*

Proof: Obviously, we need only show that the dimension of Z over P is equal to the degree m of the minimum polynomial $\varphi(x)$ of the matrix a. Let us consider an extension Ω of the field P such that a can be reduced over Ω to the

Jordan normal form. Then, in accordance with Lemma 7, the dimension over Ω of the center C of the algebra $Z\Omega_n(a)$ coincides with m. However, it is easy to show that the dimension of C over Ω is equal to the dimension of Z over P since, by virtue of Lemma 5, $ZP_n(a)$ contains a P-basis $b_1, \ldots, b_\varrho$ that is also an Ω-basis for $Z\Omega_n(a)$. Consequently, C consists of all solutions of the system

$$b_\nu x = x b_\nu, \qquad \nu = 1, \ldots, \rho, \qquad (1.37)$$

where $x \in \Omega_n$, and Z consists of all solutions of the same system, where $x \in P_n$. The system (1.37) is equivalent to a system of homogeneous linear equations in n^2 unknowns x_{ij}, where $\|x_{ij}\| = x$. This implies the coincidence of the Ω-dimension of C with the P-dimension of Z. This means that the dimension of Z over P is equal to m. This completes the proof of the theorem.[4]

Theorem 8. $ZP_n(a)$ *is a field if and only if the characteristic polynomial of the matrix a is irreducible over P.*

Proof: If the characteristic polynomial of a is irreducible, it coincides with its minimum polynomial and, on the basis of Theorem 5, $ZP_n(a) = P[a]$. But $P[a]$ is a field (by Corollary 1 of Section 1).

Conversely, if $ZP_n(a)$ is a field, the minimum polynomial of a coincides with the characteristic polynomial since, on the basis of Theorem 7, $ZP_n(a) = P[a]$. Since $P[a]$ is a field, the minimum polynomial of a is irreducible.

---

[4] An interesting generalization of Theorem 7 is given in [45].

2

Commutative Subgroups of $GL(n, P)$ and Commutative Subalgebras of P_n

1. A Connection between Two Problems

Let M denote a maximum set of pairwise-commutative matrices in P_n. Then, M is a maximal subalgebra in the class of all the commutative subalgebras of the full linear algebra P_n. Moreover, any matrix in P_n that commutes with every matrix in M is contained in M. In particular, for $a \in M$, $b \in M$, and $\lambda \in P$, we have $a + b \in M$, $ab \in M$, and $\lambda a \in M$.

Every commutative subalgebra of P_n is contained in a maximal commutative subalgebra of P_n since the rank of P_n over P is bounded.

We shall now establish a connection between the maximal commutative subalgebras P_n and the maximal commutative subgroups $GL(n, P)$. We shall assume that the field P contains at least three elements. We denote by $L = L_n$ the set of all maximal commutative subalgebras of P_n and we denote by $K = K_n$ the set of all maximal commutative subgroups of $GL(n, P)$.

Every algebra in L obviously contains the unit matrix E_n. If $A \in L$, we shall denote by A^* the group of all invertible elements of the algebra A. We define a function f on the set L by setting, for every $a \in L$,

$$f(A) = A^*. \tag{2.1}$$

We shall say that the algebra A is *decomposable* if it is possible to represent the space P^n as the direct sum of invariant subspaces with respect to A and we shall say that it is *indecomposable* otherwise.

Lemma 1.   *If the field P contains more than two elements, then $f(A) \in K$ for $a \in L$.*

Proof: Let us first consider the case in which A is indecomposable. Then, on the basis of Lemma 4 of Chapter 1, every matrix in A has no more than one eigenvalue in the field P. Let $E_n, a_1, \ldots, a_r$ denote a basis for the algebra A. Obviously, for $\lambda_j \in P$, the matrices $E_n, a_1 - \lambda_1 E_n, \ldots, a_r - \lambda_r E_n$ also belong to A and they constitute a basis for A. Let us now choose the λ_j in such a way that no λ_j is an eigenvalue of the matrix a_j. Then, $a_j - \lambda_j E_n \in A^*$. Thus, A^* contains a basis for the algebra A.

Suppose now that $c \in GL(n, P)$ and that c commutes with every matrix of the group A^*. Then, c also commutes with every matrix in A, $c \in A$, and $c \in A \cap GL(n, P) = f(A)$. Thus, $f(A) \in K$.

On the other hand, if A is indecomposable, then, in a suitable basis for the space P^n, every matrix $a \in A$ can be written in the form

$$a = [a_1, \ldots, a_k], \tag{2.2}$$

where a_j ranges over the indecomposable algebra A_j of degree n_j, for $j = 1, \ldots, k$ and $n_1 + \cdots + n_k = n$. Since the algebra A is maximal in the class of all the commutative subalgebras of P_n, the blocks a_j in (2.2) range over the algebras A_j independently of each other. Let δ denote an element of the field P other than 0 or 1. The matrices $d_j = [E_{n_1}, \ldots, \delta E_{n_j}, \ldots, E_{n_k}]$ commute with every matrix (2.2) and, consequently, $d_j \in A^*$. But then, every matrix $c \in GL(n, P)$ that commutes with every matrix of A^* has quasi-diagonal form (2.2): $c = [c_1, \ldots, c_k]$.

Since $f(A_j) \in K_{n_j}$, we have $c_j \in f(A_j)$ and $c \in f(A)$. Consequently, $f(A)$ is maximal among the commutative subgroups of $GL(n, P)$. This completes the proof of the lemma.

Let Γ denote a group in K. Then, we denote by $[\Gamma]$ the linear P subspace spanned by the set Γ.

We define a function g on the set K by setting $g(\Gamma) = [\Gamma]$ for an arbitrary group $\Gamma \in K$.

Lemma 2. *Let P denote an arbitrary field and let Γ denote an indecomposable group belonging to K. Then, $g(\Gamma) \in L$.*

Proof: Suppose that a matrix a commutes with every matrix of the algebra $[\Gamma]$. Obviously, the algebra generated by a and Γ is indecomposable. Consequently, P contains no more than one eigenvalue of the matrix a. Thus, there exists a $\lambda \in P$ such that $a - \lambda E_n$ is a nonsingular matrix. Therefore, $a - \lambda E_n \in \Gamma$, $a \in [\Gamma]$, and $[\Gamma] \in L$. This completes the proof of the lemma.

Let H denote an indecomposable maximal commutative subgroup of $GL(n, P)$. The matrices $h \in H$ can all be written in the form

$$h = \begin{bmatrix} a_{11}(h) & 0 & \ldots & \\ a_{21}(h) & a_{22}(h) & \ldots & \\ \multicolumn{4}{c}{\dotfill} \\ a_{s1}(h) & a_{s2}(h) & \ldots & a_{ss}(h) \end{bmatrix}, \qquad (2.3)$$

where $h \to a_{jj}(h)$ is an irreducible representation of H for $j = 1, \ldots, s$. We shall refer to the group D consisting of all matrices of the form

$$[a_{11}(h), a_{22}(h), \ldots, a_{ss}(h)], \qquad (2.4)$$

as the *diagonal* of the group H and we shall refer to the matrix (2.4) as the *diagonal* of the matrix (2.3).

Lemma 3. *Suppose that $\Gamma \in K_n$, that $H_1, \ldots, H_k$ are the indecomposable factors of Γ, and that $D_1, \ldots, D_k$ are the diagonals of these factors. Then, among the $D_1, \ldots, D_k$, there is at most one unit group.*

Proof: The matrices $g \in \Gamma$ can all be written

$$g = [h_1, \ldots, h_k], \qquad (2.5)$$

where the matrices h_i, for $i = 1, \ldots, k$, range independently over the groups H_i of the form (2.3). Suppose that $D_1 = (E_{n_1})$ and $D_2 = (E_{n_2})$. Then, the matrix $t = [f, E_{n-(n_1+n_2)}]$, where

$$f = \left[\begin{array}{c|c} E_{n_1} \begin{array}{ccc} 0 & \ldots & 0 & 1 \\ \multicolumn{3}{c}{\dotfill} \\ 0 & \ldots & 0 \end{array} \\ \hline 0 & E_{n_2} \end{array} \right]$$

obviously commutes with every matrix (2.5). But t does not belong to Γ, which contradicts the maximality of Γ among the commutative subgroups of $GL(n, P)$.

Lemma 4. *Let P denote an arbitrary field and suppose that $\Gamma \in K$. Then, $g(\Gamma) \in L$.*

Proof: Let $H_1, \ldots, H_k$ denote the indecomposable factors of Γ and let $D_1, \ldots, D_k$ denote their diagonals. In accordance with Lemma 3, we may assume that $D_2, \ldots, D_k$ are distinct from the unit group. Then, for $j > 1$, there exists in Γ a matrix $b = [E_n, \ldots, c_j, \ldots, E_{n_k}]$, where the diagonal of the matrix c_j is distinct from E_{n_j}. Since $c_j \in H_j$, H_j is indecomposable, and the diagonal c_j is distinct from E_{n_j}, it follows that unity is not an eigenvalue of the matrix c_j and, consequently, $c_j - E_{n_j}$ is a nonsingular matrix. Therefore, $d_j = [E_{n_1}, \ldots, g - E_{n_j}, \ldots, E_{n_k}]$ also belongs to Γ. Then,

$$q_j = (b - E_n)d_j^{-1} \in [\Gamma], \qquad j = 2, \ldots, k. \qquad (2.6)$$

Obviously, $q_j = [0_{n_1}, \ldots, E_{n_j}, \ldots, 0_{n_k}]$. Therefore, it follows from (2.6) that an arbitrary matrix $a \in P_n$ that commutes with every matrix in $[\Gamma]$ is of quasi-diagonal form $[a_1, \ldots, a_k]$. Since $g(H_j) \in L_{n_j}$, we have $a_j \in g(H_j)$. Since $q_j \in [\Gamma]$, it follows that $[\Gamma]$ is the direct sum of the algebras $[H_j]$. Consequently, $a \in [\Gamma]$ and $[\Gamma] \in L$. This completes the proof of the lemma.

Lemma 5. *Let P denote a field containing more than two elements. Then, fg is the identity mapping of K onto K and gf is the identity mapping of L onto L.*

Proof: If $\Gamma \in K$, then $fg(\Gamma) = f[\Gamma] = f(A)$, where $[\Gamma] = A \in L$. Obviously, $f(A) = A^* \supset \Gamma$. Since $\Gamma \in K$, we have $A^* = \Gamma$. Consequently, $fg(\Gamma) = \Gamma$.

Suppose now that $A \in L$. Then, $gf(A) = g(A^*) = g(\Gamma)$, where $\Gamma = A^* \in K$, and $g(\Gamma) = [\Gamma] \in L$. On the other hand, $A \supset \Gamma$. Consequently, $A \supset [\Gamma]$. Since $[\Gamma] \in L$, we have $A = [\Gamma]$ and $gf(A) = A$.

Corollary. *If the field P contains more than two elements, then*

(i) *f is a one-to-one mapping of L onto K,*

(ii) *g is a one-to-one mapping of K onto L,*

(iii) *$g = f^{-1}$.*

Proof: $f(L) = K$ since $\Gamma = fg(\Gamma) = f(A)$, where $A = g(\Gamma) \in L$, for $\Gamma \in K$. The mapping f is one-to-one since the equation $f(A) = f(A_1)$, where $A, A_1 \in L$, implies $A = gf(A) = gf(A_1) = A_1$. Result (ii) is proven analogously. Results (i) and (ii) and Lemma 5 lead to (iii).

Two algebras A and B in P_n are said to be *conjugate* in P_n if $GL(n, P)$ contains a matrix t such that $B = tAt^{-1}$.

From the definition of the mapping f [cf. formula (2.1)] and the above corollary, we get

Theorem 1. *Suppose that the field P contains more than two elements. There exists a one-to-one mapping f of the set L onto the set K such that $f(A)$ and $f(A_1)$, where A and A_1 belong to L, are conjugate in GL(n, P) if and only if A and A_1 are conjugate in P_n.*

Thus, the problem of describing all maximal commutative subalgebras of P_n is almost always equivalent to the problem of describing all maximal commutative subgroups of

$GL(n, P)$. An exception is the case in which $P = GF(2)$, where $GF(2)$ is a field with exactly two elements. If $P = GF(2)$, then, for $n > 1$, the set L contains more elements than does the set K since, in accordance with Lemma 4. $g(K) \subset L$. If $\Gamma_1 \neq \Gamma_2$ and $\Gamma_1, \Gamma_2 \in K$, then $g(\Gamma_1) \neq g(\Gamma_2)$. On the other hand, L contains the algebra D of all diagonal matrices, which does not belong to $g(K)$. In fact, if $D = g(\Gamma) = [\Gamma]$, where $\Gamma \in K$, then $\Gamma \subset D$. Consequently, $\Gamma = \{E_n\}$, which contradicts the relation $\Gamma \in K$. This means that the number of elements of L is greater than the number of elements of K.

2. General Properties of Commutative Subalgebras of the Algebra P_n

Let A denote a maximal commutative subalgebra of the full linear algebra P_n. On the basis of Theorem 3 of Chapter 1, all matrices $a \in A$ can be written simultaneously in the form

$$a = [a_1, \ldots, a_s], \tag{2.7}$$

where $a \to a_j$ is the representation of A by matrices of degree n_j, all the irreducible parts of which are equivalent. The matrices a_j range independently over the maximal commutative subalgebras of P_{n_j}, for $j = 1, \ldots, s$, where $n_1 + \cdots + n_s = n$.

Conversely, if A is the algebra of matrices of the form (2.7), where the a_i range independently over the maximal commutative subalgebras of P_{n_i}, then, obviously A contains $b_j = [0, \ldots, E_{n_j}, \ldots, 0]$ for $j = 1, \ldots, s$. Now, if $cb_j = b_j c$ for $c \in P_n$, then c is of the form (2.7) and, hence, $c \in A$. Thus, A is a maximal commutative subalgebra of P_n, and we have

Theorem 2. *Every maximal commutative subalgebra A of the full linear algebra P_n is the direct sum of its restrictions A_j (for $j = 1, \ldots, s$), each of which is the maximal commutative subalgebra of P_{n_j} with equivalent irreducible parts, $n_1 + \cdots + n_s = n$. Conversely, if A is the direct sum of the restrictions A_j and if A_j is a maximal subalgebra of the algebra P_{n_j}, where $\sum n_j = n$, then A is a maximal commutative subalgebra of P_n.*

Thus, the description of the maximal commutative subalgebras of P_n is completely reduced to the description of the maximal commutative subalgebras of P_m, for $m \leqslant n$, with equivalent irreducible parts.

It follows from Theorem 3 of Chapter 1 that, if A is an indecomposable commutative subalgebra of P_n, then all its irreducible parts are equivalent. The converse is obvious: If A is the maximal commutative subalgebra of P_n with equivalent irreducible parts, then A is indecomposable.

Now, let P denote a perfect field and let A denote a maximal commutative indecomposable subalgebra of P_n. On the basis of Theorem 4 of Chapter 1, there exists an extension Ω of the field P of degree r, where r divides $n(r/n)$, such that all the matrices of the algebra A can be written simultaneously in the form

$$a = \begin{bmatrix} \lambda & 0 & \ldots & 0 \\ \lambda_{21} & \lambda & \ldots & 0 \\ \cdots\cdots\cdots\cdots\cdots \\ \lambda_{s1} & \lambda_{s2} & \ldots & \lambda \end{bmatrix}, \qquad (2.8)$$

where λ and $\lambda_{\alpha\beta}$ belong to Ω, $s = nr^{-1}$, and $\Omega \in P_r$. Let us write the matrices (2.8) in the form

$$a = \lambda E_s + (a - \lambda E_s) = \lambda E_s + \rho. \qquad (2.9)$$

From this we get $A = \Omega E_s + R$, where R is a commutative nilpotent subalgebra of Ω_s. Since A is a maximal commutative subalgebra of Ω_s, it follows that R is a maximal commutative nilpotent subalgebra of Ω_s. Consequently, we have

Theorem 3. *Let A denote an indecomposable maximal commutative subalgebra of P_n, where P is a perfect field. Then, there exists an extension Ω of the field P of degree r, $r|n$, such that $\Omega \subset P_r$ and $A = \Omega E_s + R$, where $s = nr^{-1}$ and R is a maximal commutative nilpotent subalgebra of the algebra Ω_s.*

The converse is also true:

Theorem 4. *Let P denote an arbitrary field and let Ω denote the extension of the field P of degree r, $r|n$, where $\Omega \subset P_r$. If N is a maximal commutative nilpotent subalgebra of the algebra Ω_s, where $rs = n$, then $A = \Omega E_s + N$ is a maximal commutative subalgebra of the algebra P_n.*

Proof: Let N denote a maximal commutative subalgebra of the algebra Ω_s. Let us first establish three properties of the algebra N.

(i) *Every nilpotent matrix $U \in \Omega_s$ that commutes with every matrix belonging to N belongs to N.* This is true because $M = N \cup \{U\}$ is a set of pairwise-commutative matrices. In accordance with Theorem 2 of Chapter 1, the matrices M can be simultaneously reduced to triangular form with zero diagonal. Consequently, M generates a commutative nilpotent subalgebra N_1 of the algebra Ω_s that contains N. By virtue of the maximality of N, we have $N_1 = N$ and $U \in N$.

(ii)   *N is an indecomposable subalgebra of the algebra* Ω_s. To see this, suppose the opposite. Then, the matrices $g \in N$ can be simultaneously reduced to the form $g = [A(g), B(g)]$, where $A(g)$ and $B(g)$ are triangular matrices with zero diagonals. If the degree of $A(g)$ is equal to k, then the $(k+1)$th row and the kth column of the matrix g consist entirely of zeros. Consequently, for $g \in N$, we have $ge_{kk+1} = e_{kk+1}g = 0$. From (i), $e_{kk+1} \in N$. This last contradicts (2.8) and proves our assertion.

(iii) $A = \Omega E_s + N$ *is a maximal commutative subalgebra of the algebra* Ω_s. Let B denote a maximal commutative subalgebra of the algebra Ω_s that contains A. On the basis of (ii), B is an indecomposable subalgebra of Ω_s. Consequently, the matrices B can be simultaneously reduced to the form

$$b = \begin{bmatrix} c(b) & 0 & \ldots & 0 \\ c_{21}(b) & c(b) & \ldots & 0 \\ \multicolumn{4}{c}{\cdots\cdots\cdots\cdots\cdots\cdots\cdots} \\ c_{t1}(b) & c_{t2}(b) & \ldots & c(b) \end{bmatrix}, \qquad (2.10)$$

where $b \to c(b)$ is an irreducible representation of the algebra B of degree $m = st^{-1}$. To prove (iii), it will be sufficient for us to show that $m = 1$. Obviously, N consists of all matrices (2.10) such that $c(g) = 0$. Consequently, N consists of matrices of the form

$$\begin{bmatrix} 0 & 0 & \ldots & 0 \\ c_{21} & 0 & \ldots & 0 \\ \multicolumn{4}{c}{\cdots\cdots\cdots\cdots\cdots} \\ c_{t1} & c_{t2} & \ldots & 0 \end{bmatrix}, \qquad (2.11)$$

where $c_{\alpha\beta}$ is an $m \times m$ matrix over Ω.

Now, let d denote an arbitrary $m \times m$ matrix over Ω. Then, the matrix

$$d_1 = \begin{bmatrix} 0 & 0 & \dots & 0 \\ 0 & 0 & \dots & 0 \\ d & 0 & \dots & 0 \end{bmatrix}$$

is nilpotent and, as one can easily show, it commutes with every matrix of the form (2.11). By virtue of (i), we have $d_1 \in N$. Consequently, $bd_1 = d_1 b$, where b is an arbitrary matrix in B. In accordance with (2.10), we find from this last equation that $c(b)d = dc(b)$. Since d is an arbitrary matrix in Ω_m and the representation $b \to c(b)$ is irreducible, it follows that $m = 1$. From this (iii) follows.

From (iii), we easily obtain proof of the theorem. Suppose that $U \in P_n$ commutes with every matrix of the algebra $A = \Omega E_s + N$. Since A contains the subalgebra ΩE_s consisting of all matrices of the form $[w, \dots, w]$, where $w \in \Omega$, it follows that

$$u = \begin{bmatrix} u_{11} & \dots & u_{1s} \\ \dotfill \\ u_{s1} & \dots & u_{ss} \end{bmatrix},$$

where $u_{ij}w = wu_{ij}$. The field Ω is a maximal commutative subalgebra of P_r. Therefore, $u_{ij} \in \Omega \subset P_r$. On the basis of (iii), $u \in A$. This completes the proof of the theorem.

By virtue of Theorems 3 and 4, the problem of constructing maximal commutative subalgebras of the algebra P_n is equivalent in the case of a perfect field P to the problem of constructing maximal commutative nilpotent subalgebras of P_n. In what follows, we shall be concerned with commutative nilpotent subalgebras of the algebra P_n.

Theorem 5. *For arbitrary natural number k satisfying the inequality $2 \leqslant k \leqslant n$ and for an arbitrary field P, the set of maximal commutative nilpotent subalgebras P_n contains a subalgebra of the kth class of nilpotency.*

Proof: Suppose that $a = e_{21} + e_{32} + \cdots + e_{k\,k-1}$. Consider[1]

$$A = [a, a^2, \ldots, a^{k-1}e_{k+11}, e_{k+21}, \ldots, e_{n1}].$$

Since $a^k = 0$ and $a^{k-1} \neq 0$ and since $ae_{j1} = e_{j1}a = 0$ for $j > k$, it follows that A is a commutative nilpotent algebra of class k. Let us show that A is a maximal commutative nilpotent subalgebra of the full linear algebra P_n. Suppose that $c \in P_n$ and that c commutes with every matrix e_{j1} for $j > k$. Then,

$$c = \begin{bmatrix} \alpha_{11} & 0 & \ldots & 0 & 0 & 0 & \ldots & 0 \\ \alpha_{21} & \alpha_{22} & \ldots & \alpha_{2k} & 0 & 0 & \ldots & 0 \\ \cdots\cdots\cdots\cdots\cdots\cdots\cdots\cdots\cdots\cdots\cdots \\ \alpha_{k1} & \alpha_{k2} & \ldots & \alpha_{kk} & 0 & 0 & \ldots & 0 \\ \alpha_{k+11} & \alpha_{k+12} & \ldots & \alpha_{k+1k} & \alpha_{11} & 0 & \ldots & 0 \\ \cdots\cdots\cdots\cdots\cdots\cdots\cdots\cdots\cdots\cdots\cdots \\ \alpha_{n1} & \alpha_{n2} & \ldots & \alpha_{nk} & 0 & 0 & \ldots & \alpha_{11} \end{bmatrix}.$$

If, in addition, $ca = ac$, then

---

[1] Here, $[a, a^2, \ldots, e_{n1}]$ is the linear P-subspace spanned by the matrices $a, a^2, \ldots, e_{n1}$.

$$c = \begin{bmatrix} \alpha_{11} & 0 & 0 & \cdots & 0 & 0 & \cdots & 0 \\ \alpha_{21} & \alpha_{11} & 0 & \cdots & 0 & 0 & \cdots & 0 \\ \cdots\cdots\cdots\cdots\cdots\cdots\cdots\cdots\cdots\cdots\cdots \\ \alpha_{k1} & \alpha_{k-11} & \alpha_{k-21} & \cdots & \alpha_{11} & 0 & \cdots & 0 \\ \alpha_{k+11} & 0 & 0 & \cdots & 0 & \alpha_{11} & \cdots & 0 \\ \cdots\cdots\cdots\cdots\cdots\cdots\cdots\cdots\cdots\cdots\cdots \\ \alpha_{n1} & 0 & 0 & \cdots & 0 & 0 & \cdots & \alpha_{11} \end{bmatrix}.$$

$$(2.12)$$

If c is also nilpotent, then $\alpha_{11} = 0$. But every matrix of the form (2.12) with $\alpha_{11} = 0$ is contained in the algebra A. This completes the proof of the theorem.

Let N denote a nilpotent (not necessarily commutative) subalgebra of class k of the algebra P_n, where $2 \leqslant k \leqslant n$. Consider the subspace NP^n. We have $NP^n \neq (0)$ since $N \neq 0$. On the other hand, $NP^n \neq P^n$ since $N^{k-1}(NP^n) = N^k P^n = (0)$ but $N^{k-1} \neq (0)$. Obviously, in general, $P^n \supset NP^n \supset \cdots \supset N^{k-1}P^n \supset N^k P^n = (0)$ is a strictly descending chain of subspaces of the space P^n. Now, if

$$u_{11}, \ldots, u_{n_1 1}, u_{12}, \ldots, u_{n_2 2}, \ldots, u_{1k}, \ldots, u_{n_k k} \qquad (2.13)$$

is a basis of the space P^n such that $N^{j-1}P^n = [u_{1j}, \ldots, u_{n_j j}] \dotplus N^j P^n$, then, in the basis (2.13), all the matrices $g \in N$ are of the form

$$g = \begin{bmatrix} 0_{n_1 n_1} & 0_{n_1 n_2} & \cdots & 0_{n_1 n_k} \\ a_{n_2 n_1} & 0_{n_2 n_2} & \cdots & 0_{n_2 n_k} \\ \cdots\cdots\cdots\cdots\cdots\cdots\cdots \\ a_{n_k n_1} & a_{n_k n_2} & \cdots & 0_{n_k n_k} \end{bmatrix}, \qquad (2.14)$$

where $a_{n_i n_j}$ is an $n_i \times n_j$ matrix over P and $0_{n_i n_j}$ is the zero $n_i \times n_j$ matrix. On the other hand, since the product of

k arbitrary matrices of the form (2.14) is equal to 0, we have proven

Theorem 6.   (i)   *The matrices of an arbitrary nilpotent subalgebra of the full linear algebra P_n over an arbitrary field P can be simultaneously transformed to triangular form* (2.14).

(ii)   *An arbitrary algebra M of matrices of the form* (2.14) *is nilpotent and the class of its nilpotence does not exceed k.*

(iii)   *All maximal nilpotent subalgebras of the full linear algebra P_n are conjugate in it.*

3. Nilpotent Subalgebras of Class 2

Let P denote an arbitrary field and let N denote a nilpotent subalgebra of P_n of class 2. According to (2.14), P^n contains a basis such that the matrices of the algebra N simultaneously assume the form

$$\begin{bmatrix} 0_{n_1} & 0_{n_1 n_2} \\ a_{n_2 n_1} & 0_{n_2} \end{bmatrix}, \qquad n_1 + n_2 = n. \qquad (2.15)$$

Obviously, the set of all matrices of the form (2.15) is a subalgebra $N_{n_1 n_2}$ of dimension $n_1 n_2$ over P. Also, $N_{n_1 n_2}^2 = (0)$, and $N_{n_1 n_2}$ is commutative. Direct verification shows that an arbitrary nilpotent matrix belonging to P_n that commutes with every matrix in $N_{n_1 n_2}$ belongs to $N_{n_1 n_2}$. Thus, $N_{n_1 n_2}^2$ is a maximal commutative nilpotent subalgebra of P_n. Since $N_{n_1 n_2}^2 = 0$, the algebra $N_{m_1 m_2}$ is isomorphic to the algebra $N_{n_1 n_2}$ if and only if their dimensions over P coincide, that is, if $m_1 m_2 = n_1 n_2$. Since $n_1 + n_2 = m_1 + m_2 = n$, the isomorphism $N_{n_1 n_2}^2 \cong N_{m_1 m_2}$ holds if and only if either $n_1 = m_1$ and $n_2 = m_2$ or $n_1 = m_2$ and $n_2 = m_1$. Consequently, in P_n there are $[\frac{1}{2}n]$ isomorphic maximal commutative nilpotent subalgebras that are algebras of class 2.

However, for $n_1 \neq n_2$, isomorphic algebras $N_{n_1 n_2}$ and $N_{n_2 n_1}$ are not conjugate in P_n. To see this, suppose, for example, that $n_1 < n_2$ and that the complete linear group $GL(n, P)$ contains a matrix t such that $tN_{n_1 n_2} = N_{n_2 n_1} t$. The matrices in $tN_{n_1 n_2}$ are of the form $[A_{nn_1} 0_{nn_2}]$ and those belonging to $N_{n_2 n_1} t$ are of the form

$$\begin{bmatrix} 0_{n_2 n} \\ B_{n_1 n} \end{bmatrix}.$$

Consequently, $tN_{n_1 n_2}$ consists of matrices of the form

$$\begin{bmatrix} 0_{n_2 n_1} & 0_{n_2 n_2} \\ c_{n_1 n_1} & 0_{n_1 n_2} \end{bmatrix}.$$

Then,

$$tN_{n_1 n_2} : P \leqslant n_1{}^2 < n_1 n_2 = N_{n_1 n_2} : P.$$

This last is impossible since t is a nonsingular matrix. Thus, we have proven

Theorem 7. *The set of all maximal commutative nilpotent subalgebras of the full linear algebra P_n has exactly $n - 1$ nonconjugate algebras of class 2.*

We note that the dimension of each nilpotent subalgebra of class 2 of the algebra P_n does not exceed the number $[\frac{1}{4}n^2]$ and that, for arbitrary $n > 1$, P_n contains a nilpotent algebra of class 2, the dimension of which over P is equal to $[\frac{1}{4}n^2]$. To prove this, let us consider separately the two cases $n = 2\nu$ and $n = 2\nu + 1$.

If $n = 2\nu$, we have $[\tfrac{1}{4}n^2] = \nu^2$. Let N denote a nilpotent subalgebra of class 2 of the algebra P_n. Then, $N:P \leqslant n_1 n_2$, where $n_1 + n_2 = n$. Also,

$$[\tfrac{1}{4}n^2] - n_1 n_2 = \nu^2 - (2\nu - n_1)n_1 = \nu^2 - 2\nu n_1 + n_1{}^2$$

$$= (\nu - n_1)^2 \geqslant 0.$$

Consequently, $N:P \leqslant [\tfrac{1}{4}n^2]$ and $N_{\nu\nu}:P = [\tfrac{1}{4}n^2]$.

If $n = 2\nu + 1$, we have $[\tfrac{1}{4}n^2] = \nu^2 + \nu$. Again, let N denote a nilpotent class 2 subalgebra of the full linear algebra P_n. Then, $N:P \leqslant n_1 n_2$. But

$$[\tfrac{1}{4}n^2] - n_1 n_2 = \nu^2 + \nu - (2\nu + 1 - n_1)n_1$$

$$= \nu^2 - 2\nu n_1 + n_1{}^2 + (\nu - n_1)$$

$$= (\nu - n_1)^2 + (\nu - n_1)$$

$$= (\nu - n_1)(\nu - n_1 + 1) \geqslant 0.$$

Consequently, $N:P \leqslant [\tfrac{1}{4}n^2]$ and $N_{\nu\nu+1}:P = [\tfrac{1}{4}n^2]$.

We shall show below that the rank of every commutative nilpotent (not necessarily class 2) subalgebra of the full linear algebra P_n does not exceed $[\tfrac{1}{4}n^2]$.

4. Commutative Nilpotent Subalgebras of the Algebra P_n of Class n

Suppose that the characteristic of the field P is either equal to 0 or greater than $n - 1$ and that N is a maximal commutative nilpotent subalgebra of class n of the algebra P_n. According to Proposition 6 of Chapter 1, the algebra N contains a matrix a such that $a^{n-1} \neq 0$.

Let us reduce the matrix a to Jordan normal form: $a = e_{21} + e_{32} + \cdots + e_{n\,n-1}$. Let b denote a matrix in N. The equation $ab = ba$ implies, on the basis of Theorem 5 of Chapter 1, that

$$b = \beta_0 E_n + \beta_1 a + \beta_2 a^2 + \cdots + \beta_{n-1} a^{n-1}.$$

Since b is also nilpotent, we have $\beta_0 = 0$. Consequently, the algebra N has a basis $a, a^2, \ldots, a^{n-1}$. Thus, we have proven

Theorem 8. *Up to conjugacy in P_n, the algebra P_n possesses a unique commutative nilpotent subalgebra N of class n:*

$$N = [a, a^2, \ldots, a^{n-1}],$$

where $a = e_{21} + e_{32} + \cdots + e_{n\,n-1}$. *The algebra N is maximal among the commutative nilpotent subalgebras of the algebra P_n.*

5. Kravchuk's Normal Form

Throughout this section, P denotes an arbitrary field. Let N denote a commutative nilpotent subalgebra of the full linear algebra P_n. The set of all elements $c \in N$ that satisfy the condition $cN = (0)$ constitutes a subalgebra M of the algebra N, which we shall call the *annihilator* of the algebra N. $M \neq (0)$ because M contains, for example, N_{k-1} if N is of class k.

Kravchuk's First Theorem. *Let N denote a commutative nilpotent subalgebra of the algebra P_n of class $k > 2$.*

Suppose that N is not contained in any other commutative nilpotent subalgebra of P_n of the same class. Then, the matrices g of the algebra N can be simultaneously reduced by a similarity transformation to the form

$$g = \begin{bmatrix} 0_{\nu\nu} & 0_{\nu m} & 0_{\nu\mu} \\ a_{m\nu}(g) & b_{mm}(g) & 0_{m\mu} \\ c_{\mu\nu}(g) & d_{\mu m}(g) & 0_{\mu\mu} \end{bmatrix}, \qquad (2.16)$$

where $\nu + m + \mu = n$, $\nu, m, \mu \neq 0$. Here, the annihilator M of the algebra N consists of all matrices of the algebra P_n that are of the form

$$\begin{bmatrix} 0_{n-\mu\ \nu} & 0_{n-\mu\ n-\nu} \\ c_{\mu\nu} & 0_{\mu\ n-\nu} \end{bmatrix}. \qquad (2.17)$$

Consequently, the dimension $M:P$ is $\mu\nu$.

Proof: The subspace NP^n does not coincide with P^n and it contains nonzero vectors. Consider the subspace Q of vectors q in NP^n such that $Nq = 0$. Since $N^{k-1}P^n \leqslant Q$, it follows that $Q \neq (0)$. On the other hand, $Q \neq NP^n$ since $N^2 \neq (0)$. Let us now write a basis for the space P_n

$$u_1, \ldots, u_\nu, v_1, \ldots, v_m, q_1, \ldots, q_\mu, \qquad (2.18)$$

such that $q_1, \ldots, q_\mu$ is a basis for the subspace Q and $NP^n = [v_1, \ldots, v_m] + Q$. In the basis (2.18), the matrices of the algebra N are of the form (2.16).

We still need to show that the annihilator M of the algebra N consists of all matrices of the algebra P_n of the form (2.17). Suppose that $c \in M$. Then, $cN = (0)$, $NcP^n = (0)$, and $cP^n \leqslant Q$. Consequently, c is of the form (2.17). Now, if d is a matrix belonging to P_n of the form (2.17), then $dNP^n = 0$; that is, $dN = 0$. Furthermore, $dP^n \leqslant Q$. Consequently, $Nd = 0$. Then, $d \in M$ since N is a maximal commutative nilpotent subalgebra of P_n of class k.

In what follows, we shall refer to the form of the matrices (2.16)–(2.17) of a commutative nilpotent algebra N as Kravchuk's normal form and we shall refer to the basis (2.18) as an N-basis.

We note that the numbers ν, m, and μ in Kravchuk's form of the algebra N are uniquely defined. Specifically, $\nu = n - (NP^n : P)$ and $\nu\mu = M : P$. We shall refer to the vector (ν, m, μ) as the *Kravchuk signature* of the algebra N.

Let us add that the subspace Q constructed in the proof of the theorem coincides with the subspace of all vectors in P_n that are annihilated by every operator of the algebra N. To see this, note that $w \in NP^n$ and $Nw = 0$. In the N-basis of the space P^n, we take the vector w for u_1. Then, the first column of every matrix of N will consist only of zeros. This contradicts the arbitrariness of the matrix $c_{\mu\nu}$.

Kravchuk's Second Theorem. *Suppose that a commutative nilpotent subalgebra N of the full linear algebra P_n of class $k > 2$ consists of matrices of the form*

$$g = \begin{bmatrix} 0_{\nu\nu} & 0_{\nu m} & 0_{\nu\mu} \\ a_{m\nu}(g) & b_{mm}(g) & 0_{m\mu} \\ c_{\mu\nu}(g) & d_{\mu m}(g) & 0_{\mu\mu} \end{bmatrix}.$$

Then, $b_{mm}(g)$ ranges over a commutative nilpotent subalgebra B of the algebra P_m of class $k - 2$.

Proof: Let (2.18) denote an N-basis of the space P^n and set $Q = [q_1, \ldots, q_\mu]$. Then, $N^{k-1}P^n \subset Q$ and, consequently, $N^{k-2}[v_1, \ldots, v_m] = N^{k-1}P^n \subset Q$. Therefore, the product of any $k - 2 - x$ matrices $b_{mm}(g)$ is equal to 0; that is, $b_{mm}(g)$ ranges over the algebra B, the class of which does not exceed $k - 2$.

Suppose now that $B^{k-3} = (0)$. Then,

$$N^{k-3}[v_1, \ldots, v_m] \subset Q,$$
$$N^{k-2} P^n = N^{k-3}([v_1, \ldots, v_m] \dotplus Q) \subset Q,$$

and

$$N^{k-1} P^n \subset NQ = (0).$$

But $N^{k-1} \neq (0)$. This contradiction completes the proof of the theorem.

We now introduce a condition for maximality of a commutative nilpotent subalgebra of the algebra P_n.

Theorem 9. *Suppose that a commutative nilpotent subalgebra N of the algebra P_n consists of matrices of the form (2.16) and that its annihilator contains all matrices of the form (2.17). If $P_n \backslash N$ contains no nilpotent matrices of the form (2.16) that commute with every matrix in N, then N is a maximal commutative subalgebra of P_n.*

Proof: Suppose that

$$z = \begin{bmatrix} A_{\nu\nu} & B_{\nu m} & C_{\nu\mu} \\ D_{m\nu} & E_{mm} & F_{m\mu} \\ 0_{\mu\nu} & G_{\mu m} & H_{\mu\mu} \end{bmatrix}$$

is a matrix in the centralizer of the algebra N. Since $zc = cz$ for an arbitrary matrix c of the form (2.17), we have

$$C_{\nu\mu}c = 0, \qquad F_{m\mu}c = 0, \qquad cB_{\nu m} = 0, \qquad H_{\mu\mu}c = cA_{\nu\nu},$$

$$(2.19)$$

where c is an arbitrary $\mu \times \nu$ matrix. From (2.19), we obtain the results $C_{\nu\mu} = 0$, $F_{m\mu} = 0$, $B_{\nu m} = 0$, and $A_{\nu\nu}$ and $H_{\mu\mu}$ are scalar matrices. If the matrix z is also nilpotent, then $A_{\nu\nu} = 0$, $H_{\mu\mu} = 0$, and z is of the form (2.16). This completes the proof of the theorem.

6. Kravchuk's Third Theorem. Symmetric Signatures

Let N denote a commutative nilpotent subalgebra of the algebra P_n of class $k > 2$ and suppose that N is not contained in any other commutative nilpotent subalgebra of the algebra P_n of the same class. Suppose that (2.16)–(2.17) is the Kravchuk normal form of the matrices of N. Suppose, furthermore, that the matrices

$$n_i = \begin{bmatrix} 0_{\nu\nu} & 0_{\nu m} & 0_{\nu\mu} \\ A^i & B^i & 0_{m\mu} \\ C^i & D^i & 0_{\mu\mu} \end{bmatrix}, \qquad i = 1,\ldots,s \qquad (2.20)$$

of the form (2.16) constitute a basis for the algebra N. Let us construct an $m \times \nu s$ matrix A, writing one after the other the matrices $A^1,\ldots, A^s$:

$$A = [A^1 A^2 \cdots A^s]. \qquad (2.21)$$

Lemma 6. *The rank of the matrix A is m.*

Proof: Suppose that the rank of A is equal to r and that $r < m$. Let us show that, in this case, it is possible, without violating the form (2.16)–(2.17) of the matrices of the algebra N, to reduce them to a form such that the first $m - r$ rows of the matrix A, that is, the first $m - r$ rows of each matrix A^i in (2.20), will consist of zeros. By making elementary transformations on the rows, we can reduce A to a form such that its first $m - r$ rows consist of zeros. But this means that $GL(n, P)$ contains a matrix T such that the first $m - r$ rows of the matrix TA consist of zeros. $TA = [TA^1, \ldots, TA^s]$. Now, if $F = [E_\nu, T, E_\mu]$, then

$$Fn_iF^{-1} = \begin{bmatrix} 0_{\nu\nu} & 0_{\nu m} & 0_{\nu\mu} \\ TA^i & TB^iT^{-1} & 0_{m\mu} \\ * & D^iT^{-1} & 0_{\mu\mu} \end{bmatrix}.$$

Therefore, we can assume that, if $r < m$, the first $m - r$ rows of all the matrices $a_{m\nu}(g)$ in (2.16), for $g \in N$, consist of zeros. But

$$n_i g = \begin{bmatrix} 0_{\nu\nu} & 0_{\nu m} & 0_{\nu\mu} \\ b_{mm}(g)A^i & * & 0_{m\mu} \\ * & * & 0_{\mu\mu} \end{bmatrix}.$$

Therefore, the first $m - r$ rows of the matrices $b_{mm}(g)A_i$, for $g \in N$, also consist of zeros. Let us set

$$b_{mm}(g) = \begin{bmatrix} b_{m-r\,m-r}(g) & b_{m-r\,r}(g) \\ b_{r\,m-r}(g) & b_{rr}(g) \end{bmatrix}, \qquad A = \begin{bmatrix} 0_{m-r\,\nu s} \\ A_{r\,\nu s} \end{bmatrix}.$$

Then, we obtain

$$b_{m-r\,r}(g)A_{r\,\nu s} = 0. \qquad (2.22)$$

But the rank of the matrix $A_{r\nu s}$ is equal to r. Therefore, it follows from (2.22) that $b_{m-r\,r}(g) = 0$.

Thus, the form (2.16)–(2.17) of the matrices of the algebra N may be rewritten

$$g = \begin{bmatrix} 0_\nu & 0_{\nu m} & & 0_{\nu\mu} \\ 0_{m-r\nu} & b_{m-r\ m-r}(g) & 0_{m-r\ r} & 0_{m\mu} \\ a_{r\nu}(g) & b_{r\ m-r}(g) & b_{rr}(g) & \\ c_{\mu\nu}(g) & d_{\mu m} & & 0_{\mu\mu} \end{bmatrix}.$$

As g ranges over the algebra N, the matrix $b_{m-r\ m-r}(g)$ ranges over a commutative nilpotent subalgebra of the algebra P_{m-r}. Therefore, $GL(m-r, P)$ contains a matrix f such that all matrices of the form

$$f b_{m-r\ m-r}(g) f^{-1} = c_{m-r\ m-r}(g)$$

have a lower triangular form with zeros on the principal diagonal. Now, if $t = [E_\nu, f, E_{r+\mu}]$, then the matrices $q = tgt^{-1}$ are of the form

$$q = \begin{bmatrix} 0_{\nu\nu} & 0_{\nu m} & & 0_{\nu\mu} \\ 0_{m-r\ \nu} & C_{m-r\ m-r}(g) & 0_{m-r\ r} & \\ & \bar{b}_{r\ m-r}(g) & b_{rr}(g) & 0_{m\mu} \\ c_{\mu\nu}(g) & \bar{d}_{\mu m}(g) & & 0_{\mu\mu} \end{bmatrix}. \qquad (2.23)$$

The first $\nu + 1$ rows of the matrices (2.23) consist entirely of zeros. Therefore, for an arbitrary matrix c of the form

$$c = \begin{bmatrix} 0_{n-\mu\ \nu+1} & 0_{n-\mu\ n-\nu-1} \\ c_{\mu\ \nu+1} & 0_{\mu\ n-\nu-1} \end{bmatrix} cq = qc = 0.$$

This is impossible since the rank of the annihilator of the algebra N is equal to $\nu\mu$. This completes the proof of the lemma.

Kravchuk's Third Theorem. *Let N denote a commutative nilpotent subalgebra of the full linear algebra P_n of class $k > 2$. Suppose that N is not contained in any other commuta-*

tive nilpotent subalgebra of the algebra P_n of the same class. Suppose, furthermore, that N consists of matrices of the form (2.16)–(2.17). Then, if $a_{m\nu}(g) = 0$ for $g \in N$, we have $b_{mm}(g) = 0$ and $d_{\mu m}(g) = 0$.

Proof: Let (2.20) denote a basis for the algebra N. Let g denote a matrix in N such that $a_{m\nu}(g) = 0$. We may assume that $c_{\mu\nu}(g)$ is also zero since the matrix

$$\begin{bmatrix} 0_{n-\mu\ \nu} & 0_{n-\mu\ n-\nu} \\ c_{\mu\nu}(g) & 0_{\mu\ n-\nu} \end{bmatrix}$$

belongs to N. For $j = 1, \ldots, s$, we have $gn_j = n_jg$. Consequently, $b_{mm}(g)A^j = 0$ and $d_{\mu m}(g)A^j = 0$. But then,

$$b_{mm}(g)A = 0, \qquad d_{\mu m}(g)A = 0, \qquad (2.24)$$

where A is the matrix (2.21). The rank of the matrix A is equal to m (cf. Lemma 6). Then, it follows from (2.24) that $b_{mm}(g) = 0$ and $d_{\mu m}(g) = 0$.

Corollary. *Suppose that $g_1, \ldots, g_s$ are of the form (2.16) and belong to N. Suppose also that $c_{\mu\nu}(g_i) = 0$. The matrices $g_1, \ldots, g_s$ are linearly independent if and only if $a_{m\nu}(g_1), \ldots, a_{m\nu}(g_s)$ are linearly independent.*

Proof: Let ν, m, and μ denote natural numbers such that $\nu + m + \mu = n$. We denote by Q the set of all maximal commutative nilpotent subalgebras of P_n of class $k > 2$ that have signature (ν, m, μ). Analogously, we denote by R the set of all maximal commutative nilpotent subalgebras of P_n of class k that have signature (μ, m, ν). Then, for arbitrary k such that $2 < k \leqslant n$, we have

Theorem 10. *There exists a one-to-one mapping ψ of the set Q onto R such that, for N and N_1 in Q, ψN and ψN_1 are conjugate in P_n if and only if N and N_1 are conjugate. This mapping ψ can be chosen in such a way that, for $N \in Q$, $N' = \psi N$ is obtained from the algebra N by transposing its matrices.*

Proof: Suppose that $N \in Q$ and that N' is obtained from the algebra N by transposing its matrices. Obviously, N' is a commutative nilpotent subalgebra of the algebra P_n that is isomorphic to the algebra N. Since N is maximal among the commutative nilpotent subalgebras of P_n of class k, so is N'. Suppose now that (v, m, μ) is the signature of the algebra N and that (v', m', μ') is the signature of N'. Since N and N' are isomorphic and $v\mu$ is the dimension of the annihilator of the algebra N, we have $v\mu = v'\mu'$. Let us show that $(\mu, m, v) = (v', m', \mu')$. Obviously, it will be sufficient to show that $v = \mu'$. Let us suppose that the matrices N in the basis (2.18) are of the form (2.16)–(2.17). Let Q denote the subspace of those vectors q in P^n such that $N'q = 0$. It follows from (2.16) and (2.18) that $Q \supset [u_1, \dots, u_v]$. If $Q \neq [u_1, \dots, u_v]$, then there exists a basis $P^n: u_1, \dots, u_v, w_1, \dots, w_{n-v}$ such that $w_1 \in Q_1$. Consequently, there exists a matrix $t \in GL(n, P)$ such that the first $v + 1$ columns of each matrix of the algebra $t^{-1}N't$ consist of zeros. Then, the first $v + 1$ rows of each matrix of the algebra $t'Nt'^{-1}$ consist of zeros. But this is impossible since the signature of the algebra $t'Nt'^{-1}$ coincides with the signature of the algebra N and $v = n - (NP^n : P)$. Consequently, $Q = [u_1, \dots, u_s]$, $\mu' = v$, $v' = \mu$, and $m' = m$. This completes the proof of the theorem.

Corollary. *If the last μ rows of a matrix c of the algebra N of the form (2.16)–(2.17) consist of zeros, then $c = 0$.*

Proof: Suppose that (2.18) is the N-basis of the space P^n in which the matrices $c \in N$ are of the form (2.16)–(2.17). Then,

$$q_1, \ldots, q_\mu, v_1, \ldots, v_m, u_1, \ldots, u_\nu \qquad (2.25)$$

is an N'-basis for P^n. To see this, note that $[u_1, \ldots, u_\nu] = Q_1$ is a subspace of all vectors $q \in P^n$ such that $N'q = 0$ (cf. proof of the above theorem). Furthermore, $N'P^n \subset [v_1, \ldots, v_m, u_1, \ldots, u_\nu]$ since the matrices $c' \in N'$ are of the form

$$c' = \begin{bmatrix} 0_{\nu\nu} & a'_{m\nu} & c_{\mu\nu'} \\ 0_{m\nu} & b'_{mm} & d_{\mu m'} \\ 0_{\mu\nu} & 0_{\mu m} & 0_{\mu\mu} \end{bmatrix}$$

in the basis (2.18). By virtue of the preceding theorem, the dimension of the space $N'P^n$ is equal to $m + \nu$. Therefore, $N'P^n = [v_1, \ldots, v_m, u_1, \ldots, u_s]$. The matrices $q_1, \ldots, q_\mu$ supplement the basis $v_1, \ldots, v_m, u_1, \ldots, u_\nu$ of the subspace $N'P^n$ so as to constitute a basis for P^n. Consequently, (2.25) is an N'-basis of the space P^n. In the basis (2.25), the matrices $c' \in N'$ have the Kravchuk normal form. Also, the first μ columns of the matrix c' consist of zeros. Consequently, $c' = 0$ (by Kravchuk's third theorem) and $c = 0$.

Corollary. *Suppose that $g_1, \ldots, g_s$ are of the form (2.16) and that they belong to N. Suppose also that $c_{\mu\nu}(g_i) = 0$ for $i = 1, \ldots, s$. The matrices $g_1, \ldots, g_s$ are linearly independent, if and only if the $d_{\mu m}(g_i)$ are linearly independent.*

7. A Regular Representation of a Commutative Nilpotent Algebra

If A is a nilpotent algebra of dimension n over a field P, its regular representation is not faithful because in this

representation, the zero operator corresponds to all elements
in the annihilator of the algebra A. However, it is possible
to construct an isomorphic representation of the algebra A
by matrices of degree $n + 1$. To do this, let us put A into
the algebra B of rank $n + 1$:

$$B = Pe \dotplus A,$$

where $ea = ae = a, e^2 = e, a \in A$. The regular representation of the algebra B is faithful because B has a unit element.
In this representation, the algebra C corresponding to A
is isomorphic to it. We shall call C the regular representation
of degree $n + 1$ of the algebra A.

Theorem 11. *If A is an n-dimensional commutative
algebra with a unit element over the field P, then its regular
representation is a maximal commutative subalgebra of the
full linear algebra P_n.*

Proof: Let us first show that, if A is an algebra with
unit element over the field P, the centralizer in P_n of its
left regular representation coincides with its right regular
representation. Indeed, if f and τ are, respectively, the
operators of the left and right regular representations of the
algebra A, then, for all $x \in A$, we have $f(x) = ax$, and
$\tau(x) = xb$ for $a, b \in A$. Also, $(f\tau)(x) = f(xb) = axb = \tau(ax) =
(\tau f)x$.

Conversely, suppose that f is the operator of the left
regular representation of the algebra A and σ is an operator
that commutes with it. Then, for arbitrary $x \in A$, we have
$(\sigma f)x = \sigma(ax) = (f\sigma)x = a\sigma(x)$ for $a \in A$. If we set $x = e$,
where e is the unit element of the algebra A, we obtain
$\sigma(a) = a\sigma(e)$; that is, σ is the operator of the right regular
representation of the algebra.

If the algebra is commutative, its left and right regular representations coincide. This completes the proof of the theorem.

Theorem 12. *If A is a commutative nilpotent algebra of dimension n over P, then its regular representation of degree $n + 1$ is a maximal commutative nilpotent subalgebra of the full linear algebra P_{n+1}.*

Proof: If C is the regular representation of degree $n + 1$ of the algebra A, then $B = PE_{n+1} \dotplus C$ is the regular representation of the algebra $Pe \dotplus A$, which, by virtue of the preceding theorem, is a maximal commutative subalgebra of the algebra P_{n+1}. The matrices of the algebra C can be simultaneously reduced to triangular form with zeros on the diagonal by a similarity transformation (cf. Theorem 2, Chapter 1). Consequently, we may assume that B consists of triangular matrices with scalar diagonal. But then, B is nonsingular and, by virtue of Theorem 3, C is a maximal commutative nilpotent subalgebra of the algebra P_{n+1}.

Theorem 13. *If N is a maximal commutative nilpotent subalgebra of the algebra P_n of class $k > 2$ with signature $(1, m, \mu)$, then the dimension of N over P is equal to $n - 1$ and N is conjugate in P_n with its regular representation.*

Proof: According to Kravchuk's first theorem, the matrices g of the algebra N can be simultaneously transformed to the form

$$g = \begin{bmatrix} 0 & 0_{1m} & 0_{1\mu} \\ a_{m1}(g) & b_{mm}(g) & 0_{m\mu} \\ c_{\mu 1}(g) & d_{\mu m}(g) & 0_{\mu\mu} \end{bmatrix}.$$

Now, if $g_1, \ldots, g_s$ is a basis for N, then, by virtue of Lemma 6, the rank of the matrix $A = [a_{m1}(g_1) \ldots a_{m1}(g_s)]$ is equal to m. From this it follows that the matrix $a_{m1}(g)$ ranges over the entire space of $m \times 1$ matrices over P as g ranges over the entire algebra N. On the basis of the corollary to Kravchuk's third theorem, it then follows that the dimension of N over P is equal to $m + \mu = n - 1$. Therefore, we can choose, as a basis of N, elements $g_1, g_2, \ldots, g_{n-1}$ such that, for $i \leqslant m$,

$$a_{m1}(g_i) = \begin{bmatrix} 0 \\ \vdots \\ 1 \\ \vdots \\ 0 \end{bmatrix}, \quad 1$$

in the ith row,

$$b_{mm}(g_i) = ||\beta_{kl}^i||, \qquad c_{\mu 1}(g_i) = 0, \qquad d_{\mu m}(g_i) = ||\delta_{kl}^i||,$$

and for $i = m + j$, $j > 0$

$$a_{m1}(g_i) = 0, \qquad b_{mm}(g_i) = 0, \qquad c_{\mu i}(g_i) = \begin{bmatrix} 0 \\ \vdots \\ 1 \\ \vdots \\ 0 \end{bmatrix}, \quad 1$$

in the jth row, $d_{\mu m}(g_i) = 0$.

Let us now consider the matrices of a regular representation γ of the algebra N in the basis

$$E_n, g_1, g_2, \ldots, g_{n-1}. \tag{2.26}$$

$\gamma(g) = \sigma_g$, where $\sigma_g(x) = gx$ for $g \in N$ and $x \in N$. Let us show that the matrix of the operator σ_{g_i} in the basis (2.26) coincides with g_i, for $i = 1, \ldots, n - 1$. We have $\sigma_{g_i}(E_n) = g_i$. Consequently, the first column of σ_{g_i} coincides with the first column of g_i. Consider $\sigma_{g_i}(g_j) = g_i g_j$. If $i > m$, then $g_i g_j = 0$. Consequently, for $i > m$, we have $\sigma_{g_i} = g_i$. If $j > m$, then $g_i g_j = 0$ and, consequently, the $(j + 1)$th column of σ_{g_i}, like the $(j + 1)$th column of g_i, consists of zeros. It remains only to consider the case $i \leqslant m$, $j \leqslant m$ and to show that the $(j + 1)$th column of σ_{g_i} coincides with the $(j + 1)$th column of the matrix g_i. In this case, we have[2]

$$\sigma_{g_i}(g_j) = g_i g_j$$

$$= \begin{bmatrix} 0 & 0_{1m} & 0_{1\mu} \\ a_{m1}(g_i) & b_{mm}(g_i) & 0_{m\mu} \\ 0_{\mu 1} & d_{\mu m}(g_i) & 0_{\mu\mu} \end{bmatrix} \begin{bmatrix} 0 & 0_{1m} & 0_{1\mu} \\ a_{m1}(g_j) & b_{mm}(g_j) & 0_{m\mu} \\ 0_{\mu 1} & d_{\mu m}(g_j) & 0_{\mu\mu} \end{bmatrix}$$

$$= \begin{bmatrix} 0 & 0_{1m}0_{1\mu} \\ b_{mm}(g_i)a_{m1}(g_j)^* & 0_{m\mu} \\ d_{\mu m}(g_i)a_{m1}(g_j)^* & 0_{\mu\mu} \end{bmatrix} = \begin{bmatrix} 0 & \cdots \\ \beta_{1j}^i & \cdots \\ \beta_{2j}^i & \cdots \\ \cdot & \cdots \\ \beta_{mj}^i & \cdots \\ \delta_{1j}^i & \cdots \\ \cdot & \cdots \\ \delta_{\mu j}^i & \cdots \end{bmatrix}$$

$$= \sum_{k=1}^{m} \beta_{kj}^i g_k + \sum_{l=1}^{\mu} \delta_{lj}^i g_{m+l}.$$

---

[2] On the basis of Kravchuk's third theorem, the matrix of the algebra N with signature $(1, m, \mu)$ is uniquely determined by its first column.

Thus, the $(j + 1)$th column of the matrix σ_{g_i} coincides with the column

$$\begin{bmatrix} 0 \\ \beta^i_{1j} \\ \dots \\ \beta^i_{mj} \\ \delta^j_{1j} \\ \dots \\ \delta^i_{\mu j} \end{bmatrix}. \qquad (2.27)$$

However, (2.27) is the $(j + 1)$th column of the matrix g_i. Thus, $\sigma_{g_i} = g_i$. This completes the proof of the theorem.

Theorem 14. *Let A denote a commutative nilpotent algebra of dimension $n - 1$ over a field P. Then, its regular representation of degree n is a maximal commutative nilpotent subalgebra of the algebra P_n with signature $(1, m, n - m - 1)$.*

Proof: Let N denote the regular representation of the algebra A of degree n and let $P^n = PE_n \dotplus N$ denote the space of the representation. Then, $NP^n = N(PE_n \dotplus N) = N = P^{n-1}$.

A consequence of the two preceding theorems is

Theorem 15. *A maximal commutative nilpotent subalgebra of the full linear algebra P_n is conjugate with its regular representation of degree n if and only if the first number of its Kravchuk signature is 1.*

Corollary 1. *Let N_1 and N_2 denote two maximal commutative nilpotent subalgebras of the algebra P_n, the first* *number of the signature of each of which is* 1. *Then, isomorphism of these algebras implies their conjugacy in* P_n.

Corollary 2. *Let N_1 and N_2 denote two maximal commutative nilpotent subalgebras of the algebra P_n, the last number of the signature of each of which is* 1. *Then, isomorphism of these algebras implies their conjugacy in* P_n.

8. Commutative Nilpotent Subalgebras of Class 3 of the Full Linear Algebra P_n

We shall now show that if the field P is infinite, the set of all maximal commutative nilpotent subalgebras of the full linear algebra P_n is, for $n > 6$, an infinite set of pairwise-nonconjugate and even nonisomorphic subalgebras of class 3.

Let N denote a commutative nilpotent algebra of class 3 over an arbitrary field P and let r denote the dimension of N. Let us denote by M the annihilator of the algebra N. If $v_1, \ldots, v_t$ is a basis for M, we write a basis for the algebra N in the form

$$u_1, \ldots, u_s, v_1, \ldots, v_t, s + t = r. \qquad (2.28)$$

Since N is of class 3, we have $N^2 \subset M$. Therefore,

$$u_i u_j = \sum_{k=1}^{t} \alpha_{ij}^k v_k,$$

where $\alpha_{ij}^k = \alpha_{ji}^k \in P$. It is easy to see that the description of commutative nilpotent algebras of class 3 reduces to the case in which the annihilator M of the algebra N coincides with N^2. The basis for the algebra N can be chosen in the form $u_1, \ldots, u_s,$ $v_1, \ldots, v_\nu,$ $w_1, \ldots, w_\mu,$ where $v_1, \ldots, v_\nu,$

$w_1, \ldots, w_\mu$ is a basis for M and $v_1, \ldots, v_\nu$ is a basis for N^2. Let us set $N_1 = [u_1, \ldots, u_s, v_1, \ldots, v_\nu]$. Then, N is the direct sum $N = N_1 + W$, where $W = [w_1, \ldots, w_\mu]$, $W^2 = (0)$, $WN_1 = (0)$.

Up to isomorphism, the algebra N is given by $s^2 t$ numbers α_{ij}^ν or t symmetric matrices $A^{(\nu)} = ((\alpha_{ij}^\nu))$ of degree s. We shall call the matrices $A^{(\nu)}$ the *structural matrices* of the algebra N. It is easy to determine how the structural matrices of the algebra N change as we shift from the basis (2.28) to another basis

$$a_1, \ldots, a_s, b_1, \ldots, b_t \tag{2.29}$$

of the same form (where $b_1, \ldots, b_t$ is the basis M). Suppose that

$$a_i = \sum_\nu \gamma_{\nu i} u_\nu + v^i, \quad v_i \in M, \quad v^i = \sum_\mu \beta_{\mu i} b_\mu, \quad \gamma_{\nu i}, \beta_{\mu i} \in P.$$

We set

$$C = ||\gamma_{\nu i}||, \quad B = ||\beta_{\mu i}||, \quad a_i a_j = \sum_\rho \delta_{ij} b_\rho, \quad ||\delta_{ij}|| = D^{(\rho)},$$

where the $D^{(\rho)}$ are the structural matrices in the basis (2.29). Then,

$$
\begin{aligned}
a_i a_j &= \left(\sum_\nu \gamma_{\nu i} u_\nu \right) \left(\sum_\mu \gamma_{\mu j} u_\mu \right) \\
&= \sum_{\mu\nu} \gamma_{\nu i} \gamma_{\mu j} \sum_k \alpha_{\nu\mu}^k v_k \\
&= \sum_{\mu,\nu} \gamma_{\nu i} \gamma_{\mu j} \sum_k \alpha_{\nu\mu}^k \sum_\rho \beta_{\rho k} b_\rho \\
&= \sum_\rho \delta_{ij}^\rho b_\rho, \quad \delta_{ji}^\rho
\end{aligned}
$$

$$= \sum_{\mu,\nu} \gamma_{\nu i}\gamma_{\mu j} \sum_k \alpha_{\nu\mu}^k \beta_{\rho k}$$

$$= \sum_k \beta_{\rho k} \sum_{\nu,\mu} \gamma_{\nu i}\alpha_{\nu\mu}^k \gamma_{\mu j}.$$

This means that

$$D^{(\rho)} = \sum_{k=1}^t \beta_{\rho k} C' A^{(k)} C, \qquad \rho = 1,\ldots,s, \qquad (2.30)$$

where C' is the transpose of C.

Suppose now that $M = N^2$. Then, the matrices $A^{(1)},\ldots,A^{(t)}$ are linearly independent. Let A denote the space generated by the matrices $A^{(1)},\ldots,A^{(t)}$ and let R denote the set of ranks of the matrices in A. It follows from formulas (2.30) that R does not change when we shift from one basis of the algebra N of the form (2.28) to another basis of the same form.

Consider the t-dimensional space of bilinear forms in s pairs of variables over the field P with basis $\varphi_1,\ldots,\varphi_t$, where φ_ν is a form with matrix $A^{(\nu)}$. If a basis

$$\varphi = \begin{bmatrix} \varphi_1 \\ \cdots \\ \varphi_t \end{bmatrix}$$

for the space of forms is replaced with a new basis

$$\psi = \begin{bmatrix} \psi_1 \\ \cdots \\ \psi_t \end{bmatrix},$$

where $\psi = B\varphi$, and then the variables of the forms are subjected to a linear transformation by the matrix C, then the matrices of the forms of the basis obtained for the new space are expressed in terms of $A^{(1)}, \ldots, A^{(\nu)}$ with the aid of formulas (2.30).

We shall say that two spaces of bilinear forms over the field P are equivalent if there exists a linear nonsingular transformation with coefficients in P that maps one of the spaces into the other.

It follows from (2.30) that two commutative nilpotent algebras of class 3 (the case $M = N^2$) are isomorphic if and only if the corresponding spaces of bilinear forms are equivalent.

Corollary 3. *If* $\frac{1}{2}s(s + 1) = t$, *there exists* (*up to isomorphism*) *exactly one commutative nilpotent algebra N of class 3 over the field P such that the dimension $N:P = s + t$, $N^2:P = t$, and the annihilator $M = N^2$.*

Corollary 4. *If* $M = N^2$, *then* $t \leqslant \frac{1}{2}s(s + 1)$.

Let us now prove that there are infinitely many non-isomorphic nilpotent commutative algebras of class 3 of dimension 6 over an infinite field P.

We shall denote by N_α a commutative nilpotent algebra whose structural matrices are of the form $A^{(1)} = E_4$, $A^{(2)} = A_\alpha = [-1, 0, 1, \alpha]$, where $\alpha \in P$. Then, in the notation of this section, we shall have: $s = 4, t = 2, u_1^2 = v_1 - v_2$, $u_2^2 = v_1$, $u_3^2 = v_1 + v_2$, and $u_4^2 = v_1 + \alpha v_2$. For $i \neq j$, we have $u_i u_j = u_j u_i = 0$ and $u_i v_j = v_j u_i = v_j v_i = 0$.

Lemma 7. *Let α denote a field element of the field P such that* $(\alpha^3 - \alpha)(\alpha^2 - \frac{1}{9}) \neq 0$. *Then, P contains only finitely many elements β such that the algebras N_α and N_β are isomorphic.*

Proof: Suppose that $N_\alpha \cong N_\beta$. Then, on the basis of (2.30), there exist in $GL(n, P)$ matrices C and $B = ||\beta_{ij}||$ such that

$$E_4 = \beta_{11}C'E_4C + \beta_{12}C'A_\alpha C,$$
$$A_\beta = \beta_{21}C'E_4C + \beta_{22}C'A_\alpha C. \qquad (2.31)$$

From this it follows that $C'C = C'E_4C$ and $C'A_\alpha C$ are diagonal matrices. The relations (2.31) are not violated if instead of C we take λC and instead of B we take $\lambda^{-2}B$, where λ is a nonzero element of P. Here, we may assume that $C'C = d = [1, \delta_1, \delta_2, \delta_3]$, where δ_j is a nonzero element of P. Therefore, $C' = dC^{-1}$ and $C'A_\alpha C = dC^{-1}A_\alpha C$, where $C^{-1}A_\alpha C$ is a diagonal matrix with the same eigenvalues as A_α. We denote by $\alpha_1, \alpha_2, \alpha_3$, an arbitrary permutation of the numbers $-1, 1, \alpha$. Then, four cases are possible: (1) $C^{-1}A_\alpha C = [0, \alpha_1, \alpha_2, \alpha_3]$; (2) $C^{-1}A_\alpha C = [\alpha_1, 0, \alpha_2, \alpha_3]$; (3) $C^{-1}A_\alpha C = [\alpha_1, \alpha_2, 0, \alpha_3]$; and (4) $C^{-1}A_\alpha C = [\alpha_1, \alpha_2, \alpha_3, 0]$.

Let us consider case (1). Then, Eqs. (2.31) take the form

$$E_4 = \beta_{11}d + \beta_{12}d[0, \alpha_1, \alpha_2, \alpha_3],$$
$$A_\beta = \beta_{21}d + \beta_{22}d[0, \alpha_1, \alpha_2, \alpha_3],$$

which is equivalent to the system of equations

$$\begin{array}{ll}
\beta_{11} = 1 & \beta_{21} = -1 \\
\beta_{11}\delta_1 + \beta_{12}\alpha_1\delta_1 = 1 & \beta_{21}\delta_1 + \beta_{22}\alpha_1\delta_1 = 0 \\
\beta_{11}\delta_2 + \beta_{12}\alpha_2\delta_2 = 1 & \beta_{21}\delta_2 + \beta_{22}\alpha_2\delta_2 = 1 \\
\beta_{11}\delta_3 + \beta_{12}\alpha_3\delta_3 = 1 & \beta_{21}\delta_3 + \beta_{22}\alpha_3\delta_3 = \beta
\end{array} \qquad (2.32)$$

Let us rewrite this system as follows:

$$\delta_1 + \beta_{12}\alpha_1\delta_1 = 1 \qquad -1 + \beta_{22}\alpha_1 = 0 \qquad \beta_{11} = 1$$

$$\delta_2 + \beta_{12}\alpha_2\delta_2 = 1 \qquad -\delta_2 + \beta_{22}\alpha_2\delta_2 = 1 \qquad \beta_{21} = -1$$

$$\delta_3 + \beta_{12}\alpha_3\delta_3 = 1 \qquad -\delta_3 + \beta_{22}\alpha_3\delta_3 = \beta \qquad (2.33)$$

From (2.33), we obtain $\beta_{22} = 1/\alpha_{11}$. Furthermore, $\alpha_2 \neq \alpha_1$ and $\alpha_3 \neq \alpha_1$ since $\alpha^3 - \alpha = 0$ for $\alpha = \pm 1$. Therefore,

$$\delta_2 = \frac{\alpha_1}{\alpha_2 - \alpha_1}, \qquad \delta_3 = \frac{\alpha_1\beta}{\alpha_3 - \alpha_1},$$

and

$$\beta_{12} = \frac{\alpha_2 - 2\alpha_1}{\alpha_1\alpha_2} = \frac{\alpha_3 - \alpha_1 - \alpha_1\beta}{\alpha_1\alpha_3\beta},$$

so that

$$\beta = \frac{\alpha_2(\alpha_3 - \alpha_1)}{\alpha_2\alpha_3 - 2\alpha_1\alpha_3 + \alpha_1\alpha_2},$$

and

$$\delta_1 = \frac{\alpha_2}{2(\alpha_2 - \alpha_1)}.$$

It is easy to verify that the denominator in the expression for β does not vanish if we replace α_1, α_2, α_3 with an arbitrary permutation of the numbers 1, -1, α provided that α satisfies the condition of the lemma. Consequently, corresponding to α are no more than 6 (that is, 3!) values of β such that $N_\alpha \cong N_\beta$.

Cases (2)–(4) are proven in an analogous manner. This completes the proof of the lemma.

Theorem 16. *For arbitrary $r > 5$ and an infinite field P, there are infinitely many nonisomorphic commutative nilpotent algebras of class 3 of dimension r over P.*

Proof: For $r = 6$, the theorem follows from Lemma 7. If $r = 6 + k$, let us construct the direct sum $R_\alpha = N_\alpha \dotplus M$, where M is a zero algebra of rank k: $N_\alpha M = M N_\alpha = M^2 = (0)$. It is easy to see that $R_\alpha \cong R_\beta$ if and only if $N_\alpha \cong N_\beta$. The theorem then follows from Lemma 7.

Theorem 17. *If $n > 6$ and the field P is infinite, then there are infinitely many nonisomorphic maximal commutative nilpotent subalgebras of class 3 of the algebra P_n.*

Proof: If for each algebra P_α (cf. the preceding theorem) of dimension $n - 1$, we construct its regular representation of degree n, we obtain an infinite sequence of nonisomorphic maximal commutative nilpotent subalgebras of the full linear algebra P_n.

9. Commutative Nilpotent Algebras of Dimension 5

In Chapter 3, we shall need a classification of commutative nilpotent algebras of class 3 of dimension 5 over the field of complex numbers P. Below, we shall make such a classification. Here, we shall basically follow the article by Z. M. Dyment [10]. As was noted in Section 8, to describe all the commutative nilpotent algebras of class 3 of dimension r, it will be sufficient to describe all commutative algebras N such that N^2 coincides with the annihilator of N and the dimension of N does not exceed r. Suppose now that N is a commutative nilpotent algebra of class 3 of dimension 5 over the field of complex numbers and suppose that its annihilator M coincides with N^2. Just as in Section 8, we write a basis for N in the form

$$u_1, \ldots, u_s, v_1, \ldots, v_t, \tag{2.34}$$

where $s + t = 5$ and $v_1, \ldots, v_t$ is a basis for M. Since $t \leqslant \frac{1}{2}s(s + 1)$ [cf. Section 8, Corollary 2], three cases are possible: $t = 2$, $t = 1$, and $t = 3$.

Let us first consider the case $t = 2$. In this case, in accordance with Section 8, the algebra N is defined by two structural matrices A_1 and A_2, where A_1 and A_2 are linearly independent symmetric complex 3×3 matrices. In Section 8, it was shown how the structural matrices change when we switch from a basis for the algebra N of the form (2.34) to another basis of the same form. Let us denote by A the subspace generated by the matrices A_1 and A_2.

Lemma 8. *A contains matrices of rank exceeding* 1.

Proof: Suppose that the rank of no matrix in A exceeds 1. Then, in accordance with Section 8, we can assume that $A_1 = [1, 0, 0]$ and that

$$A_2 = \begin{bmatrix} 0 & \alpha & \beta \\ \alpha & \gamma & \delta \\ \beta & \delta & \mu \end{bmatrix}.$$

All the second-order minors of the matrix $xA_1 + A_2$, where $x \in P$, are equal to 0; i.e., $x\gamma - \alpha^2 = 0$, $x\delta - \alpha\beta = 0$, and $x\mu - \beta^2 = 0$. Therefore, $\gamma = \alpha = \delta = \mu = \beta = 0$ and $A_2 = 0$. This last is impossible since $M = N^2$. This completes the proof of the lemma.

Let us consider separately the two possibilities: (a) there are nonsingular matrices in the space A; and (b) all the matrices in A are singular.

(a) Suppose that there are nonsingular matrices in A. Then, we may assume that $A_1 = E_3$. For A_2, we may choose the normal form of a complex symmetric matrix (cf. [7]). Consequently, three subcases are possible here:

$$(1) \qquad A_2 = [\lambda_1, \lambda_2, \lambda_3];$$

$$(2) \qquad A_2 = \begin{bmatrix} \lambda_1 & 0 & 0 \\ 0 & \lambda_2 + i & 1 \\ 0 & 1 & \lambda_2 - i \end{bmatrix};$$

$$(3) \qquad A_2 = \begin{bmatrix} \lambda & 1 & 0 \\ 1 & \lambda & -i \\ 0 & -i & \lambda \end{bmatrix};$$

(cf. [7]).

(1) Suppose that $A_2 = [\lambda_1, \lambda_2, \lambda_3]$. Since $A_1 = E_3$ and since A_1 and A_2 are linearly independent, it follows that A_2 is a nonscalar matrix. Consequently, there are two possibilities: (α) $\lambda_1 = \lambda_2 \neq \lambda_3$; and ($\beta$) all the λ_i are distinct.

For possibility (α), we may set $A_2 = [0, 0, 1]$ so that instead of A_2, we can take

$$(A_2 - \lambda_1 E_3)(\lambda_3 - \lambda_1)^{-1} = [0, 0, 1].$$

Then,

$$N = [u_1, u_2, u_3, v_1, v_2], \qquad (2.35)$$

where $u_1{}^2 = u_2{}^2 = v_1$, $u_3{}^2 = v_1 + v_2$, and all the other products of the basis elements are equal to 0. We denote this algebra by N_1.

For possibility (β), we can set $A_2 = [0, \lambda, 1]$, where $\lambda^2 \neq \lambda$. Thus, we obtain an algebra $N = [u_1, u_2, u_3, v_1, v_2]$, where $u_1{}^2 = v_1$, $u_2{}^2 = v_1 + \lambda v_2$, $u_3{}^2 = v_1 + v_2$, and the remaining products of the basis elements are equal to 0. Let us now eliminate the parameter λ. If in N we choose a new basis a_1, a_2, a_3, b_1, b_2 such that

$$a_1 = \sqrt{\lambda - 1}\, u_1,\; a_2 = u_2,\; a_3 = \sqrt{\tfrac{1}{2}\lambda},$$

$$v_1 = (\lambda - 1)^{-1}b_1,\; v_2 = [(2/\lambda) - (\lambda - 1)^{-1}]b + (2/\lambda)b_2,$$

we obtain $a_1{}^2 = b_1$, $a_2{}^2 = b_1 + 2b_2$, and $a_3{}^2 = b_1 + b_2$. Consequently, we may assume that N is defined by structural matrices $A_1 = E_3$ and $A_2 = [0, 2, 1]$. We denote by N_2 the algebra N with basis u_1, u_2, u_3, v_1, v_2 and with structural matrices E_3 and $[0, 2, 1]$.

(2) Suppose now that

$$A_2 = \begin{bmatrix} \lambda_1 & 0 & 0 \\ 0 & \lambda_2 + i & 1 \\ 0 & 1 & \lambda_2 - i \end{bmatrix}.$$

Then, A_2 can be reduced either to the form

$$\begin{bmatrix} 0 & 0 & 0 \\ 0 & i & 1 \\ 0 & 1 & -i \end{bmatrix} \tag{2.36}$$

or to the form

$$\begin{bmatrix} 1 & 0 & 0 \\ 0 & i & 1 \\ 0 & 1 & -i \end{bmatrix}. \tag{2.37}$$

This is true because, if $\lambda_1 = \lambda_2$, then, by taking $A_2 - \lambda_1 E_3$ instead of A_2, we obtain the matrix (2.36). On the other hand, if $\lambda_1 \neq \lambda_2$, we can first reduce A_2 to the form

$$\begin{bmatrix} \lambda & 0 & 0 \\ 0 & i & 1 \\ 0 & 1 & -i \end{bmatrix},$$

where $\lambda \neq 0$. As one can easily show, the two polynomial matrices

$$F(x) = xE_3 + \begin{bmatrix} 1 & 0 & 0 \\ 0 & i & 1 \\ 0 & 1 & -i \end{bmatrix}$$

where $\lambda \neq 0$,

$$G(x) = x\lambda E_3 + \begin{bmatrix} \lambda & 0 & 0 \\ 0 & i & 1 \\ 0 & 1 & -i \end{bmatrix},$$

are equivalent. Consequently, on the basis of familiar theorems in matrix theory [35], there exists a matrix T such that simultaneously $T'\lambda ET = E$ and

$$T' \begin{bmatrix} \lambda & 0 & 0 \\ 0 & i & 1 \\ 0 & i & -i \end{bmatrix} T = \begin{bmatrix} 1 & 0 & 0 \\ 0 & i & 1 \\ 0 & 1 & -i \end{bmatrix},$$

where T' is the transpose of T. From this it follows that, in case (2), we obtain two algebras: N_3 with structural matrices E_3 (2.36); and N_4 with structural matrices E_3 (2.37).

(3) The third case, in which

$$A_2 = \begin{bmatrix} \lambda & 1 & 0 \\ 1 & \lambda & -i \\ 0 & -i & \lambda \end{bmatrix},$$

obviously determines an algebra N_5 with structural matrices

$$E_3, \begin{bmatrix} 0 & 1 & 0 \\ 1 & 0 & -i \\ 0 & -i & 0 \end{bmatrix}.$$

Thus, we have proven

Lemma 9. *Case (a) defines five algebras:* $N_1, N_2, N_3, N_4,$ *and* N_5.

Lemma 10. *The algebras* N_1–N_5 *are pairwise-non-isomorphic.*

Proof: For each of the algebras N_1–N_5, let us write the sets of ranks of matrices in the set A. These will be the following sets of numbers: For N_1, there will be numbers 3, 2, and 1; for N_2, there will be 3, 2; for N_3, there will be 3, 1; for N_4, there will be 3, 2; for N_5, there will be 3, 2. From this it is obvious that it will be sufficient for us to show that the algebras N_2, N_4, and N_5 are pairwise-non-isomorphic.

Let us first show that N_4 and N_5 are not isomorphic to N_2. We set $B = [0, 2, 1]$ (the structural matrix of the algebra N_2),

$$A = \begin{bmatrix} 1 & 0 & 0 \\ 0 & i & 1 \\ 0 & 1 & -i \end{bmatrix}$$

for the algebra N_4, and

$$A = \begin{bmatrix} 0 & 1 & 0 \\ 1 & 0 & -i \\ 0 & -i & 0 \end{bmatrix}$$

for the algebra N_5 (the structural matrices of these algebras). Then, if $N_4 \cong N_2$ or $N_5 \cong N_2$, it follows from the results of Section 8 that there exist two matrices $T = ||\beta_{ij}|| \in GL(2,P)$ and $C \in GL(3, P)$ such that

$$\beta_{11}C'EC + \beta_{12}C'AC = E, \qquad \beta_{21}C'EC + \beta_{22}C'AC = B.$$

$$(2.38)$$

The determinant $|\beta_{ij}| \neq 0$, and B and E are diagonal matrices. Therefore, it follows from (2.38) that $C'EC = D$ and $C'AC$ are also diagonal matrices. Furthermore, $C' = DC^{-1}$ and $C'AC = DC^{-1}AC$; consequently, $C^{-1}AC$ is a diagonal matrix. Therefore, it is impossible for the matrix A not to be similar to a diagonal matrix. Therefore, $N_4 \ncong N_2$ and $N_5 \ncong N_2$, where the symbol $\ncong$ means that the preceding and following expressions are not isomorphic to each other.

It still remains to show that $N_4 \ncong N_5$. Let us suppose, to the contrary, that $N_5 \cong N_4$. We set

$$A = \begin{bmatrix} 0 & 1 & 0 \\ 1 & 0 & -i \\ 0 & -i & 0 \end{bmatrix}, \qquad B = \begin{bmatrix} 1 & 0 & 0 \\ 0 & i & 1 \\ 0 & 1 & -i \end{bmatrix}.$$

Then, there are two matrices $T = \|\beta_{ij}\| \in GL(2, P)$ and $C \in GL(3, P)$ that satisfy Eqs. (2.38). It follows from (2.38) that

$$D = C'EC = \begin{bmatrix} \alpha & 0 & 0 \\ 0 & \gamma_{11} & \gamma_{12} \\ 0 & \gamma_{12} & \gamma_{22} \end{bmatrix}$$

and

$$C'AC = \begin{bmatrix} \delta & 0 & 0 \\ 0 & \mu_{11} & \mu_{12} \\ 0 & \mu_{12} & \mu_{22} \end{bmatrix}.$$

But $C'AC = DC^{-1}AC$ and, consequently,

$$C'AC = \begin{bmatrix} \nu & 0 & 0 \\ 0 & \varepsilon_{11} & \varepsilon_{12} \\ 0 & \varepsilon_{12} & \varepsilon_{22} \end{bmatrix}.$$

Since this last is impossible, the lemma is proven.

Thus, we have completed the study of case (a).

(b) Suppose now that every matrix of the subspace A is singular. Then, in accordance with Lemma 8, we can assume that one of the structural matrices of the algebra N is of rank 2. Consequently, we can set $A_1 = [1, 1, 0]$. Suppose that

$$A_1 = [1, 1, 0] \quad \text{and} \quad A_2 = \begin{bmatrix} a & b & c \\ b & d & e \\ c & e & f \end{bmatrix}$$

are the structural matrices of the algebra N. Now, if $T = [T_1, 1]$, where T_1 is a complex orthogonal 2×2 matrix, then

$$T'A_1T = A_1$$

and

$$T'A_2T = \begin{bmatrix} T_1' \begin{bmatrix} ab \\ bd \end{bmatrix} T_1 & T_1' \begin{bmatrix} c \\ e \end{bmatrix} \\ [ce]T_1 & f \end{bmatrix}.$$

Consequently, without changing A_1, we can reduce the matrix

$$\begin{bmatrix} a & b \\ b & d \end{bmatrix}$$

to the normal form of a complex symmetric matrix (cf. [7]); that is, we may assume that

$$\begin{bmatrix} a & b \\ b & d \end{bmatrix}$$

is equal either to $[\alpha, \beta]$ or to

$$\begin{bmatrix} \lambda + i & 1 \\ 1 & \lambda - i \end{bmatrix}.$$

Thus, there are two possibilities for the matrix A_2:

$$A_2 = \begin{bmatrix} a & 0 & c \\ 0 & b & d \\ c & d & e \end{bmatrix} \quad \text{or} \quad A_2 = \begin{bmatrix} i & 1 & a \\ 1 & -i & b \\ a & b & c \end{bmatrix}.$$

Let us first consider the algebra N_6 defined by the structural matrices

$$A_1 = [1, 1, 0] \quad \text{and} \quad A_2 = \begin{bmatrix} a & 0 & c \\ 0 & b & d \\ c & d & e \end{bmatrix}.$$

By hypothesis, every matrix in the space A is singular. Consequently, the polynomial

$$|xA_1 + A_2| = ex^2 + [e(a + b) - (c^2 + d^2)]x \\ + abc - bc^2 - ad^2$$

is identically equal to 0. Therefore, $e = c^2 + d^2 = bc^2 + ad^2 = 0$. Furthermore, $c \neq 0$ since we would otherwise have $\alpha = 0$ and consequently u_3 would be contained in the annihilator M of the algebra N, which is impossible. Thus, $c \neq 0$, $d = \pm ic$, and $b = a$; that is,

$$A_2 = \begin{bmatrix} a & 0 & c \\ 0 & a & \pm ic \\ c & \pm ic & 0 \end{bmatrix}.$$

We may assume that

$$A_2 = \begin{bmatrix} 0 & 0 & 1 \\ 0 & 0 & \pm i \\ 1 & \pm i & 0 \end{bmatrix}$$

[since otherwise, we would only need to replace A_2 with the matrix $c^{-1}(A_2 - aA_1)$]. Consequently, the algebra N_6 possesses a basis u_1, u_2, u_3, v_1, v_2 such that $u_1{}^2 = v_1$, $u_2{}^2 = v_1$, $u_1 u_3 = v_2$, $u_2 u_3 = \pm iv_2$, and $u_3{}^2 = 0$. If, instead of u_2, we take $u_2' = -u_2$, we obtain $u_2' u_3 = iv_2$. Consequently, the algebra N_6 is defined by the structural matrices

$$A_1 = [1, 1, 0] \quad \text{and} \quad A_2 = \begin{bmatrix} 0 & 0 & 1 \\ 0 & 0 & i \\ 1 & i & 0 \end{bmatrix}.$$

Now, let us look at the algebra N defined by the structural matrices

$$A_1 = \begin{bmatrix} 1 & 0 & 0 \\ 0 & 1 & 0 \\ 0 & 0 & 0 \end{bmatrix} \quad \text{and} \quad A_2 = \begin{bmatrix} i & 1 & a \\ 1 & -i & b \\ a & b & c \end{bmatrix}.$$

By hypothesis, the expression

$$|xA_1 + A_2| = cx^2 - (a^2 + b^2)x + 2ab - b^2i + a^2i$$

is identically equal to 0. Therefore, $c = 0$, $a^2 = -b^2$, $2ab - 2b^2i = 0$, and $b \neq 0$ since otherwise u_3 would belong to the annihilator of the algebra N, which is impossible. Consequently, $a = bi$ and, for A_2, we may take

$$A_2 = \begin{bmatrix} 0 & 1 & bi \\ 1 & -2i & b \\ bi & b & 0 \end{bmatrix}.$$

The parameter b can easily be replaced by 1 since, if we take $b^{-1}u_3$ instead of u_3 in N, we obtain (instead of A_2)

$$\begin{bmatrix} 0 & 1 & i \\ 1 & -2i & 1 \\ i & 1 & 0 \end{bmatrix}.$$

Let us now show that the algebra N defined by the structural matrices

$$[1, 1, 0] \quad \text{and} \quad \begin{bmatrix} 0 & 1 & i \\ 1 & -2i & 1 \\ i & 1 & 0 \end{bmatrix} \qquad (2.39)$$

is isomorphic to the algebra N_6. Let u_1, u_2, u_3, v_1, v_2 denote the basis for N corresponding to the structural matrices (2.39). Let us choose a new basis a_1, a_2, a_3, b_1, b_2 by setting $a_1 = u_2 + iu_3$, $a_2 = u_1$, $a_3 = iu_1 + u_2$, $b_1 = v_1$, and $b_2 = v_1 - iv_2$. Then,

$$a_1{}^2 = u_2{}^2 + 2iu_2u_3 - u_3{}^2 = v_1 - 2iv_2 + 2iv_2 = v_1 = b_1;$$

$$a_1a_2 = u_2u_1 + iu_3u_1 = v_2 + i\,iv_2 = 0;$$

$$a_1a_3 = u_2{}^2 + iu_1u_2 + iu_2u_3 - u_3u_1$$
$$= v_1 - 2iv_2 + iv_2 + iv_2 - iv_2$$
$$= v_1 - iv_2 = b_2;$$

$$a_2{}^2 = u_1{}^2 = v_1 = b_1;$$

$$a_2a_3 = u_1(iu_1 + u_2) = iu_1{}^2 + u_1u_2 = iv_1 + v_2 = i(v_1 - iv_2)$$
$$= ib_2;$$

$$a_3{}^2 = -u_1{}^2 + u_2{}^2 + 2iu_1u_2 - v_1 + v_1 - 2iv_2 + 2iv_2 = 0.$$

From this follows the isomorphism $N \cong N_6$.

Thus, case (b) provides the unique algebra N_6 defined by the structural matrices

$$A_1 = [1, 1, 0] \quad \text{and} \quad A_2 = \begin{bmatrix} 0 & 0 & 1 \\ 0 & 0 & i \\ 1 & i & 0 \end{bmatrix}.$$

Obviously, N_6 is not isomorphic to any of the algebras N_1–N_5 because the ranks of the matrices in the space A are distinct for them.

Let us turn to the case $t = 1$. According to Section 8, the algebra N has in this case a single structural matrix A, which can be reduced to the form $A = [\alpha_1, \alpha_2, \alpha_3, \alpha_4]$, where $\alpha_i = 0, 1$. Consequently, $N = [u_1, u_2, u_3, u_4, v]$, where $u_i u_j = 0$ for $i \neq j$, and $u_i{}^2 = \alpha_i v$. If $\alpha_i = 0$, then u_i is contained in the annihilator of the algebra N, which contradicts the equation $t = 1$. Therefore, $\alpha_i = 1$ and $A = E_4$. We denote by N_7 the algebra defined by the single structural matrix E_4. Now, finally, suppose that $t = 3$. Then, for a basis for the algebra N, we can take

$$u_1, u_2, v_1, v_2, v_3, \qquad \text{where} \qquad u_1{}^2 = v_1, \quad u_1 u_2 = v_2,$$
$$u_2{}^2 = v_3, \quad u_i v_j = v_i v_j = 0. \tag{2.40}$$

The algebra with the basis (2.40) has the three structural matrices $[1, 0]$,

$$\begin{bmatrix} 0 & 1 \\ 1 & 0 \end{bmatrix},$$

and $[0, 1]$. We denote it by N_8.

This completes the study of the case in which $M = N^2$. All the results that we have obtained here can be summarized in

Lemma 11. *In the set of all commutative nilpotent algebras of Class 3 of dimension 5 over the field of complex numbers whose annihilators coincide with the squares of these algebras, there are only eight pairwise-nonisomorphic algebras.*

Proof: To enumerate the algebras whose annihilators do not coincide with their squares, let us first write all the algebras of dimension $r \leqslant 5$, the annihilators of which coincide with the squares of these algebras. Let us consider separately the three cases: $r = 2$; $r = 3$; and $r = 4$.

If $r = 2$, then $N = [u, v]$, where $u^2 = v$, and $uv = v^2 = 0$.

If $r = 3$, then $N = [u_1, u_2, v]$. N has only a single structural matrix A, which, in accordance with Section 8, can be reduced to the form $A = E_2$.

Suppose now that $r = 4$. Two subcases are possible: (a) The dimension of the annihilator $M : P = 1$; and (b) $M : P = 2$. In case (a), we obtain a single algebra $N = [u_1, u_2, u_3, v]$, where $u_i^2 = v$, $u_i u_j = 0$ for $i \neq j$, and $u_i v = v^2 = 0$.

Let us suppose now that case (b) obtains and that A_1 and A_2 are the structural matrices of the algebra N. Then, in the subspace $A = [A_1, A_2]$, there exists a nonsingular matrix. To see this, suppose that all the matrices of A are singular. Then, we may assume that

$$A_1 = [1, 0] \quad \text{and} \quad A_2 = \begin{bmatrix} 0 & \alpha \\ \alpha & \gamma \end{bmatrix}.$$

Since $|A_2| = 0$, we have $\alpha = 0$. Furthermore, $|A_1 + A_2| = 0$. Consequently, $\gamma = 0$, which is impossible since A_1 and A_2 are linearly independent.

Thus, we may assume that $A_1 = E_2$ and that A_2 has the normal form of a complex symmetric matrix:

$$A_2 = [0, 1] \quad \text{or} \quad A_2 = \begin{bmatrix} i & 1 \\ 1 & -i \end{bmatrix}.$$

Thus, in the case $r = 4$, we obtain three algebras.

We have noted already in Section 8 that, if N is a commutative nilpotent algebra of class 3, if M is the annihilator of the algebra N, and if $M \neq N^2$, then $N = N_1 \dotplus W$, where N_1 is a commutative nilpotent algebra, the annihilator of which coincides with its square: $NW = W^2 = (0)$. Therefore, there are only five pairwise-nonisomorphic nilpotent commutative algebras the squares of which do not coincide with their annihilators:

$$N_9 = [u, v_1, v_2, v_3, v_4],$$

where $u^2 = v_1$ and $uv_1 = v_i v_j = 0$;

$$N_{10} = [u_1, u_2, v_1, v_2, v_3],$$

where $u_1^2 = u_3^2 = v_1$ and all the other products are equal to 0;

$$N_{11} = [u_1, u_2, u_3, v_1, v_2],$$

where $u_1^2 = u_2^2 = u_3^2 = v_1$ and all the other products are equal to 0;

$$N_{12} = [u_1, u_2, v_1, v_2, v_3],$$

where $u_1^2 = v_1$, $u_2^2 = v_1 + v_2$, and all the other products are equal to 0;

$$N_{13} = [u_1, u_2, v_1, v_2, v_3],$$

where $u_1^2 = v_1 + iv_2$, $u_1 u_2 = v_2$, $u_2^2 = v_1 - iv_2$, and all the other products are equal to 0.

Thus, we have

Theorem 18. *The set of all commutative nilpotent algebras of Class 3 of dimension 5 over the field of complex* *numbers contains only* **13** *algebras that are distinct up to isomorphism:*

$$N_1 = [u_1, u_2, u_3, v_1, v_2], \quad u_1{}^2 = v_1, \quad u_2{}^2 = v_1, \quad u_3{}^2 = v_1 + v_2,$$

and the other products of the basis elements are equal to 0;

$$N_2 = [u_1, u_2, u_3, v_1, v_2], \quad u_1{}^2 = v_1, \quad u_2{}^2 = v_1 + 2v_2,$$
$$u_3{}^2 = v_1 + v_2,$$

and the other products of the basis elements are equal to 0;

$$N_3 = [u_1, u_2, u_3, v_1, v_2], \quad u_1{}^2 = v_1, \quad u_2{}^2 = v_1 + iv_2,$$
$$u_2 u_3 = v_2, \quad u_3{}^2 = v_1 - iv_2,$$

and the other products of the basis elements are equal to 0;

$$N_4 = [u_1, u_2, u_3, v_1, v_2], \quad u_1{}^2 = v_1 + v_2, \quad u_2{}^2 = v_1 + iv_2,$$
$$u_2 u_3 = v_2, \quad u_3{}^2 = v_1 - iv_2;$$
$$N_5 = [u_1, u_2, u_3, v_1, v_2], \quad u_1{}^2 = v_1, \quad u_1 u_2 = v_2,$$
$$u_2{}^2 = v_1, \quad u_2 u_3 = -iv_2, \quad u_3{}^2 = v_1,$$

and the other products of the basis elements are equal to 0;

$$N_6 = [u_1, u_2, u_3, v_1, v_2], \quad u_1{}^2 = v_1, \quad u_1 u_3 = v_2,$$
$$u_2{}^2 = v_1, \quad u_2 u_3 = iv_2,$$

and the other products of the basis elements are equal to 0;

$$N_7 = [u_1, u_2, u_3, u_4, v], \quad u_i{}^2 = v, \quad i = 1, 2, 3, 4,$$

and the other products of the basis elements are equal to 0;

$$N_8 = [u_1, u_2, v_1, v_2, v_3], \quad u_1{}^2 = v_1, \quad u_1 u_2 = v_2, \quad u_2{}^2 = v_3,$$
$$v_i v_j = 0, \quad i, j = 1, 2, 3;$$
$$N_9 = [u, v_1, v_2, v_3, v_4], \quad u^2 = v_1,$$

and the other products of the basis elements are equal to 0;

$$N_{10} = [u_1, u_2, v_1, v_2, v_3], \quad u_1{}^2 = u_2{}^2 = v_1,$$

and the other products of the basis elements are equal to 0;

$$N_{11} = [u_1, u_2, u_3, v_1, v_2], \quad u_i{}^2 = v_1, \quad i = 1, 2, 3,$$

and the other products of the basis elements are equal to 0;

$$N_{12} = [u_1, u_2, v_1, v_2, v_3], \quad u_1{}^2 = v_1, \quad u_2{}^2 = v_1 + v_2,$$

and the other products of the basis elements are equal to 0;

$$N_{13} = [u_1, u_2, v_1, v_2, v_3], \quad u_1{}^2 = v_1 + iv_2, \quad u_1 u_2 = v_2,$$
$$u_2{}^2 = v_1 - iv_2,$$

and the other products of the basis elements are equal to 0.

10. The Dimension of a Commutative Algebra of Matrices. Schur's Theorem

Theorem 19. *The dimension of a commutative nilpotent subalgebra of P_n does not exceed $[\frac{1}{4}n^2]$.*

Proof: Let N denote a maximal commutative nilpotent subalgebra of class k of the algebra P_n. We may assume that $k > 2$ since this theorem was proven for $k = 2$ in Section 3. Let us write the matrices of the algebra N in Kravchuk's form (2.16)–(2.17). A basis for the algebra N can be chosen in the form $p_1, \ldots, p_{\mu\nu}, n_1, \ldots, n_s$, where $p_1, \ldots, p_{\mu\nu}$ is a basis for the annihilator of the algebra N and $n_1, \ldots, n_s$ are matrices of the form (2.19). According to Kravchuk's third theorem, the matrices $n_1, \ldots, n_s$ are linearly independent if and only if their blocks $A^1, \ldots, A^s$ are linearly independent. Therefore, $s \leqslant m\nu$ and the dimension $N : P \leqslant \mu\nu + m\nu = (\mu + m)\nu = (n - \nu)\nu$. In Section 3, we showed that $(n - \nu)\nu \leqslant [\frac{1}{4}n^2]$. This completes the proof of the theorem.

* * *

Kravchuk ([19], [22]) assumed that we always have

$$M : P = \mu\nu + m. \qquad (2.41)$$

However, Eq. (2.41) is not always valid. An example is constructed in [28] in which $N : P > \mu\nu + m$. Recently, another counterexample, more unexpected, was discovered and is exhibited in [24] in which $N : P < \mu\nu + m$. Below, we shall consider these two examples. Suppose that N consists of all the $3\nu \times 3\nu$ matrices of the form

$$g = \begin{bmatrix} 0_{\nu\nu} & 0_{\nu\nu} & 0_{\nu\nu} \\ a_{\nu\nu}(g) & 0_{\nu\nu} & 0_{\nu\nu} \\ c_{\nu\nu}(g) & a_{\nu\nu}(g) & 0_{\nu\nu} \end{bmatrix}, \qquad (2.42)$$

where $a_{\nu\nu}(g)$ ranges over a maximal commutative subalgebra $[\frac{1}{4}\nu^2] + 1$ of an algebra P_ν of rank as g ranges over N (such subalgebras of P_ν exist, for example, $A \dot{+} PE_\nu$, where A is the maximal nilpotent commutative subalgebra of the algebra P_ν of Class 2 of rank $[\frac{1}{4}\nu^2]$) and $c_{\nu\nu}(g)$ ranges over P_ν. Then, $N:P = \nu^2 + [\frac{1}{4}\nu^2] + 1 = \rho$.

N is a maximal commutative subalgebra of the algebra P_n. Indeed, suppose that the nilpotent matrix

$$a = \begin{bmatrix} 0_{\nu\nu} & 0_{\nu\nu} & 0_{\nu\nu} \\ \alpha_{21} & \alpha_{22} & 0_{\nu\nu} \\ 0_{\nu\nu} & \alpha_{32} & 0_{\nu\nu} \end{bmatrix}$$

commutes with every matrix in N. The matrix

$$c = \begin{bmatrix} 0_{\nu\nu} & 0_{\nu\nu} & 0_{\nu\nu} \\ E_\nu & 0_{\nu\nu} & 0_{\nu\nu} \\ 0_{\nu\nu} & E_{\nu\nu} & 0_{\nu\nu} \end{bmatrix}$$

belongs to N. From the condition $ac = ca$, we obtain $\alpha_{22} = 0$ and $\alpha_{21} = \alpha_{32}$. Consequently, a is of the form

$$a = \begin{bmatrix} 0_{\nu\nu} & 0_{\nu\nu} & 0_{\nu\nu} \\ a_{\nu\nu} & 0_{\nu\nu} & 0_{\nu\nu} \\ 0_{\nu\nu} & a_{\nu\nu} & 0_{\nu\nu} \end{bmatrix},$$

where $a_{\nu\nu}$ commutes with every matrix $a_{\nu\nu}(g)$ in (2.42). Consequently, $a_{\nu\nu} = a_{\nu\nu}(g_0)$ for $g_0 \in N$ and $a = g_0$. In accordance with Theorem 9, N is a maximal commutative nilpotent subalgebra of the full linear algebra P_n.

It would follow from formula (2.41) that $N:P = \nu^2 + \nu$. But $\rho = \nu^2 + \frac{1}{4}\nu^2 + 1$ for $\nu > 3$ greater than $\nu^2 + \nu$.

Let us now look at the other counterexample.

Suppose that N is a subalgebra of the algebra P_{14} consisting of all matrices of the form

$$\begin{bmatrix} 0_{22} & 0_{2\,10} & 0_{22} \\ \begin{matrix} a\,0 \\ 0\,a \\ b\,0 \\ 0\,b \\ c\,d \\ e\,0 \\ 0\,e \\ f\,0 \\ 0\,f \\ g\,h \end{matrix} & 0_{10\,10} & 0_{10\,2} \\ c_{22} & \begin{matrix} c\,d\,0\,0\,a\,g\,h\,0\,0\,e \\ 0\,0\,c\,d\,b\,0\,0\,g\,h\,f \end{matrix} & 0_{22} \end{bmatrix}. \qquad (2.43)$$

where a, b, c, d, e, f, g, and h are arbitrary elements of P and where c_{22} is an arbitrary 2×2 matrix. Obviously, N is a commutative nilpotent subalgebra of P_{14}, $N:P = 12$. Let us show that N is a maximal commutative nilpotent subalgebra of P_{14}. Suppose that

$$z = \begin{bmatrix} 0_{22} & 0_{2\,10} & 0_{22} \\ \alpha & \gamma & 0_{10\,2} \\ 0_{22} & \beta & 0_{22} \end{bmatrix}$$

is a matrix belonging to the centralizer of the algebra N. For an arbitrary matrix

$$n_i = \begin{bmatrix} 0_{22} & 0_{2\,10} & 0_{22} \\ a_i & 0_{10\,10} & 0_{10\,2} \\ 0_{22} & b_i & 0_{22} \end{bmatrix} \in N$$

from the equation $n_i z = z n_i$, we obtain

$$\gamma a_i = 0, \qquad b_i \alpha = \beta a_i, \qquad b_i \gamma = 0. \qquad (2.44)$$

Suppose now that n_i ranges, together with four matrices of the form

$$\begin{bmatrix} 0_{12\,2} & 0_{12\,12} \\ c_{22} & 0_{2\,12} \end{bmatrix}$$

over a basis of the algebra N. Let us construct a 10×16 matrix $A = [a_1, a_2, \ldots, a_8]$, writing successively the columns of the matrices a_i, for $i = 1, 2, \ldots, 8$. By virtue of the form (2.43) of the matrices of the algebra N, it follows that the rank of A is equal to 10. On the basis of (2.44), $\gamma A = 0$. Therefore, $\gamma = 0$ and, from (2.44) there remains only the condition

$$b_i \alpha = \beta a_i. \tag{2.45}$$

Let us set $\beta = \|\beta_{ij}\|$, for $i = 1, 2$ and $j = 1, 2, \ldots, 10$ and let us set $\alpha = \|\alpha_{ij}\|$, for $i = 1, 2, \ldots, 10$ and $j = 1, 2$. Obviously, we may take

$$\alpha_{11} = \alpha_{31} = \alpha_{51} = \alpha_{55} = \alpha_{61} = \alpha_{81} = \alpha_{10\,1} = \alpha_{10\,2} = 0, \tag{2.46}$$

for which it will be sufficient to add to z a suitable linear combination of the matrices n_i.

For a_i, let us take

$$\begin{bmatrix} E_2 \\ 0_{82} \end{bmatrix}.$$

Then, $b_1 = [0_{24}\,{}^1_0\,0_{25}]$, and we obtain from (2.45), (2.46)

$$\beta_{11} = \beta_{12} = \beta_{21} = \beta_{22} = 0. \tag{2.47}$$

Suppose that

$$a_2 = \begin{bmatrix} 0_{22} \\ E_2 \\ 0_{62} \end{bmatrix}.$$

Then,

$$b_2 = \begin{bmatrix} 0_{24} & {}^0_1 & 0_{25} \end{bmatrix}$$

and from (2.45), (2.46),

$$\beta_{13} = \beta_{14} = \beta_{23} = \beta_{24} = 0. \tag{2.48}$$

For

$$a_3 = \begin{bmatrix} 0_{42} \\ 1 \ 0 \\ 0_{52} \end{bmatrix}, \qquad b_3 = \begin{bmatrix} 1 & 0 & 0 \\ 0 & 0 & 1 \end{bmatrix} 0_{27} \end{bmatrix}$$

and from (2.45), (2.46),

$$\alpha_{12} = \alpha_{32} = \beta_{15} = \beta_{25} = 0. \tag{2.49}$$

For

$$a_4 = \begin{bmatrix} 0_{42} \\ 0 \ 1 \\ 0_{52} \end{bmatrix}, \qquad b_4 = \begin{bmatrix} 0 & 1 & 0 & 0 \\ 0 & 0 & 0 & 1 \end{bmatrix} 0_{26} \end{bmatrix}$$

and from (2.45), (2.46),

$$\alpha_{21} = \alpha_{22} = \alpha_{41} = \alpha_{24} = 0. \tag{2.50}$$

For

$$a_5 = \begin{bmatrix} 0_{52} \\ E_2 \\ 0_{32} \end{bmatrix}, \qquad b_5 = [0_{29} \quad {}_0^1]$$

and from (2.45), (2.46),

$$\beta_{16} = \beta_{26} = \beta_{17} = \beta_{27} = 0. \qquad (2.51)$$

For

$$a_6 = \begin{bmatrix} 0_{72} \\ E_2 \\ 0\ 0 \end{bmatrix}, \qquad b_6 = [0_{29} \quad {}_1^0]$$

and from (2.45), (2.46),

$$\beta_{18} = \beta_{28} = \beta_{19} = \beta_{29} = 0. \qquad (2.52)$$

For

$$a_7 = \begin{bmatrix} 0_{92} \\ 1\ 0 \end{bmatrix}, \qquad b_7 = \begin{bmatrix} 0_{25} & 1 & 0 & 0 \\ & 0 & 0 & 1 \end{bmatrix} \ 0_{22}$$

and from (2.45), (2.46),

$$\alpha_{62} = \alpha_{82} = \beta_{1\,10} = \beta_{2\,10} = 0. \qquad (2.53)$$

Finally, for

$$a_8 = \begin{bmatrix} 0_{92} \\ 0\ 1 \end{bmatrix}, \qquad b_8 = \begin{bmatrix} 0_{26} & 1 & 0 & 0 & 0 \\ & 0 & 0 & 1 & 0 \end{bmatrix}$$

and from (2.45), (2.46),

$$\alpha_{71} = \alpha_{72} = \alpha_{91} = \alpha_{92} = 0. \tag{2.54}$$

It now follows from (2.47)–(2.54) that $z = 0$. On the basis of Theorem 9, N is a maximal commutative nilpotent subalgebra of P_{14}. For the algebra N, we have $\mu\nu + m = 4 + 10 > N:P$.

$$* \quad * \quad *$$

Theorem *(Schur).* *The dimension of a commutative subalgebra of the full linear algebra P_n over an arbitrary field P does not exceed $[\frac{1}{4}n^2] + 1$.*

Proof: Let us note first that, in the proof of the theorem, we can replace the field P with an arbitrary extension of it. To see this, suppose that the field Ω is an extension of the field P. Obviously, the linear space of matrices P_n is contained in the linear space of the matrices Ω_n. Now, if the matrices $l_1, \ldots, l_t$ in P_n are linearly independent over P, they are also linearly independent over Ω. Te see this, let us supplement $l_1, \ldots, l_t$ to get a basis for P_n:

$$l_1, \ldots, l_t, l_{t+1}, \ldots, l_{n2}. \tag{2.55}$$

Then, the basis of P_n

$$e_{11}, \ldots, e_{1n}, \ldots, e_{nn} \tag{2.56}$$

can be expressed as a linear combination of the elements (2.55). Obviously, (2.56) is a basis for Ω_n. Consequently, (2.55) is also a basis for Ω_n. This means that $l_1, \ldots, l_t$ are linearly independent over Ω. Now, if P_n contains ρ pairwise-commutative linearly independent matrices, then Ω_n also contains ρ pairwise-commutative linearly independent matrices.

On the basis of this, we can prove the theorem for the case in which the base field P is algebraically closed. Thus, let P denote an algebraically closed field and let A denote a maximal commutative subalgebra of P_n. Then, all irreducible parts of A will be of degree 1 (cf. Corollary 5 to Schur's Lemma, Chapter 1, Section 1). First, let us suppose that all irreducible parts of A are equivalent. Then, the matrices A can be simultaneously reduced to the form $g = \lambda(g)E_n + b(g)$, where $\lambda(g) \in P$ and b is a nilpotent triangular matrix. In other words, $A = PE_n \dotplus N$, where N is a commutative nilpotent subalgebra of P_n. According to Theorem 18, $N:P \leqslant [\frac{1}{4}n^2]$. Consequently, $A:P \leqslant [\frac{1}{4}n^2] + 1$.

Suppose now that A is a decomposable algebra. Then, all its matrices a can be simultaneously reduced by a similarity transformation to the form $a = [a_1, \ldots, a_k]$, where $k > 1$ and a_i ranges over a commutative subalgebra of the algebra P_{n_i} with equivalent irreducible parts as a ranges over the algebra A, $\sum n_i = n$ (by Theorem 2).

First, let us take the case $n = 2$. Then, $k = 2$ and all matrices of the algebra A are of the form $[\alpha_1, \alpha_2]$, where $\alpha_1, \alpha_2 \in P$, $A:P \leqslant 2 = [\frac{1}{4}2^2] + 1$.

Suppose now that $n > 2$ and that the theorem is valid for matrices of order less than n. Let us represent the matrices of the algebra A in the form $a = [a_1, b]$, where a_1 ranges over the commutative indecomposable algebra A_1 of $n_1 \times n_1$ matrices and suppose that m is the degree of b. We denote by B the subalgebra of the algebra P_m over which b ranges as a ranges over the algebra A. Then,

$$A:P \leqslant A_1:P + B:P \leqslant [\tfrac{1}{4}n_1{}^2] + 1 + [\tfrac{1}{4}m^2] + 1.$$

But

$$[\tfrac{1}{4}n^2] + 1 = [\tfrac{1}{4}(n_1{}^2 + m^2 + 2n_1 m)] + 1$$
$$\geqslant [\tfrac{1}{4}n_1{}^2] + [\tfrac{1}{4}m^2] + [\tfrac{1}{2}n_1 m] + 1$$
$$\geqslant [\tfrac{1}{4}n_1{}^2] + [\tfrac{1}{4}m^2] + 2$$

since $n > 2$ and hence $n_1 m \geqslant 2$. This completes the proof of the theorem.

The bound $[\tfrac{1}{4}n^2] + 1$ is attainable (cf. Section **3**).

3

Commutative Nilpotent Algebras of Matrices over the Field of Complex Numbers

1. Commutative Nilpotent Subalgebras of Class $n - 1$ of the Full Linear Algebra P_n

Here, we shall enumerate all the distinct (up to conjugacy) maximal commutative nilpotent subalgebras of class $n - 1$ of an algebra P_n over the complex-number field P.

Since commutative nilpotent subalgebras of the algebra of class 2 of P_3 were described in Section 3 of the preceding chapter, let us now suppose that $n > 3$ and let N denote one of these subalgebras. Then, N contains a matrix a such that $a^{n-1} = 0$ and $a^{n-2} \neq 0$ (cf. Proposition 6, Chapter 1). Let us reduce the matrix a to Jordan normal form: $a = e_{21} + e_{32} + \cdots + e_{n-1\,n-2}$. An arbitrary matrix b in P_n that commutes with a is of the form

$$
b = \begin{bmatrix}
\alpha_0 & 0 & \ldots & 0 & 0 & 0 \\
\alpha_1 & \alpha_0 & \ldots & 0 & 0 & 0 \\
\cdots\cdots\cdots\cdots\cdots\cdots\cdots\cdots \\
\alpha_{n-3} & \alpha_{n-4} & \ldots & \alpha_0 & 0 & 0 \\
\alpha_{n-2} & \alpha_{n-3} & \ldots & \alpha_1 & \alpha_0 & \beta \\
\gamma & 0 & \ldots & 0 & 0 & \delta
\end{bmatrix}.
$$

If b is nilpotent, then $\alpha_0 = \delta = 0$. Consequently, $b \, (\in N)$ can be written in the form $b = \alpha_1 a + \alpha_2 a^2 + \cdots + \alpha_{n-2} a^{n-2} + \beta e_{n-1\,n} + \gamma e_{n1}$. Therefore, a basis of the algebra N consists of the matrices $a, a^2, \ldots, a^{n-2}, \beta e_{n-1\,n} + \gamma e_{n1}, \beta_1 e_{n-1\,n} + \gamma_1 e_{n1}, \ldots$. However, a basis for the algebra N cannot contain more than a single element of the form $b = \beta e_{n-1\,n} + \gamma e_{n1}$ since, if $b_1 = \beta_1 e_{n-1\,n} + \gamma_1 e_{n1}$ belonged to N, the condition $bb_1 = b_1 b$ would imply that $\beta \gamma_1 = \beta_1 \gamma$, that is, that b and b_1 are linearly dependent. On the other hand, if N is a maximal commutative nilpotent subalgebra of the full linear algebra P_n, then one element of the form $\beta e_{n-1\,n} + \gamma e_{n1}$ must belong to a basis of the algebra N. Consequently, $N = [a, a^2, \ldots, a^{n-2}, \beta e_{n-1\,n} + \gamma e_{n1}]$ and $N : P = n - 1$.

Consider the subspace NP^n. Obviously, $NP^n \supset [u_2, \ldots, u_{n-1}]$, where $u_1, \ldots, u_n$ is that basis of the space P^n in which the matrices a and b are written. If $q = \lambda u_1 + \mu u_n$, where $\lambda, \mu \in P$, then $b(q) = \lambda \gamma u_n + \mu \beta u_{n-1}$. From this it follows that $NP^n = [u_2, \ldots, u_{n-1}]$ for $\gamma = 0$ and $NP^n = [u_2, \ldots, u_n]$ for $\gamma \neq 0$. Thus, in Kravchuk's normal form for the matrices of the algebra N, we have $\nu = 1$ for $\gamma \neq 0$ and $\nu = 2$ for $\gamma = 0$.

Let us now turn to the annihilator M of the algebra N. If $m \in M$, then, obviously, $m = \lambda a^{n-2} + \mu b$ for $\lambda, \mu \in P$, and $b = \beta e_{n-1\,n} + \gamma e_{n1}$. Also, $am = ma = 0$, $a^{n-2} b = b a^{n-2} = 0$, and $b^2 = b\gamma e_{n-1\,1} = \beta \gamma a^{n-2}$. Consequently, $M = [a^{n-2}, b]$ if at least one of the numbers β, γ is zero, and $M = [a^{n-2}]$ if $\beta, \gamma \neq 0$. Therefore, in Kravchuk's normal form for the matrices of the algebra N, we have $\mu \nu = 2$ if $\beta = 0$ or $\gamma = 0$ and $\mu \nu = 1$ for β and γ both nonzero.

Let us now consider separately the following three possibilities: (1) $\beta = 0, \gamma \neq 0$; (2) $\beta \neq 0, \gamma = 0$; (3) $\beta \neq 0, \gamma \neq 0$.

In the first case,

$$N = [a, a^2, \ldots, a^{n-2}, e_{n\,1}], \tag{3.1}$$

where $\mu\nu = 2$, $\nu = 1$, $\mu = 2$, $m = n - 3$, and $(1, n - 3, 2)$ is the signature of N.

In the second case,

$$N = [a, a^2, \ldots, a^{n-2}, e_{n-1\,n}], \qquad (3.2)$$

where $\mu\nu = 2$, $\nu = 2$, $\mu = 1$, $m = n - 3$, and $(2, n - 3, 1)$ is the signature of N.

In the third case,

$$N_\alpha = [a, a^2, \ldots, a^{n-2}, e_{n-1\,n} + \alpha e_{n\,1}], \qquad \alpha \neq 0, \quad (3.3)$$

where $\mu\nu = 1$, $\mu = \nu = 1$, $m = n - 2$, and $(1, n - 2, 1)$ is the signature of N.

The algebras (3.1)–(3.3) are not conjugate in P_n since their signatures are distinct. Let us show now that the algebras N_α shown in (3.3) are conjugate for distinct α. On the basis of Corollary 1 of Theorem 13 of Chapter 2, it will be sufficient for us to prove that these algebras are isomorphic. To do this, let us choose in N_α a basis $a, a^2, \ldots, a^{n-2}$, $(e_{n1} + \alpha e_{n-1\,n})/\sqrt{\alpha}$. In this basis, the structure constants of the algebra N_α are independent of $\alpha \neq 0$. Consequently, the algebras N_α are isomorphic for distinct α.

Theorem 1. *If $n > 3$ and P is the field of complex numbers, then every maximal commutative nilpotent subalgebra of Class $n - 1$ of the full linear algebra P_n is conjugate in P_n with one of the following three pairwise-nonconjugate algebras:*

$$N_1 = [a, a^2, \ldots, a^{n-2}, e_{n\,1}],$$
$$N_2 = [a, a^2, \ldots, a^{n-2}, e_{n-1\,n}],$$
$$N_3 = [a, a^2, \ldots, a^{n-2}, e_{n-1\,n} + e_{n\,1}],$$

where $a = e_{21} + e_{32} + \cdots + e_{n-1\,n-2}.$

2. Commutative Nilpotent Subalgebras of Class $n - 2$ of P_n

Let P denote the field of complex numbers and let N denote a maximal nilpotent commutative subalgebra of P_n whose class of nilpotency is $n - 2$. Let us assume that $n > 4$ (since N is of class 2 for $n = 4$ and all the algebras of class 2 have been described above). According to Proposition 6 of Chapter 1, N contains a matrix a such that

$$a^{n-3} \neq 0, \qquad a^{n-2} = 0. \tag{3.4}$$

Here, two cases are possible:

(a) The matrix N contains a matrix a of rank $n - 2$ satisfying condition (3.4);

(b) the algebra N does not contain a matrix a of rank $n - 2$ that satisfies condition (3.4).

Let us first consider case (a). Suppose that a matrix a of rank $n - 2$ belongs to N and satisfies condition (3.4). Obviously, $e_{21} + e_{32} + \cdots + e_{n-2\,n-3} + e_{n\,n-1}$ is the Jordan normal form of the matrix a. Consequently, we may assume that $a = e_{21} + e_{32} + \cdots + e_{n-2\,n-3} + e_{n\,n-1}$. Suppose that $b \in N$. Then, by Theorem 6 of Chapter 1, from the condition $ab = ba$, we obtain

$$b = \begin{bmatrix} \alpha_0 & 0 & \cdots & 0 & 0 & 0 \\ \alpha_1 & \alpha_0 & \cdots & 0 & 0 & 0 \\ \alpha_2 & \alpha_1 & \cdots & 0 & 0 & 0 \\ \cdots & \cdots & \cdots & \cdots & \cdots & \cdots \\ \alpha_{n-4} & \alpha_{n-5} & \cdots & 0 & \gamma_1 & 0 \\ \alpha_{n-3} & \alpha_{n-4} & \cdots & \alpha_0 & \gamma_2 & \gamma_1 \\ \beta_1 & 0 & \cdots & 0 & \delta_0 & 0 \\ \beta_2 & \beta_1 & \cdots & 0 & \delta_1 & \delta_0 \end{bmatrix}.$$

Since b is nilpotent, the eigenvalues of b and the trace of b are both zero. Therefore, $\alpha_0 = \delta_0 = 0$. Obviously, the matrix

$$b_1 = b - \alpha_1 a - \alpha_2 a^2 - \cdots - \alpha_{n-3} a^{n-3} \in N.$$

$$b_1 = \beta_1(e_{n-1\,1} + e_{n\,2}) + \beta_2 e_{n\,1} \tag{3.5}$$
$$+ \gamma_1(e_{n-3\,n-1} + e_{n-2\,n}) + \gamma_2 e_{n-2\,n-1} + \delta e_{n\,n-1}.$$

Here, we again have two possibilities: (a, α) $e_{n\,n-1} \in N$ and (a, β) $e_{n\,n-1} \notin N$.

Let us first suppose that case (a, α) occurs. Then, it follows from the condition $e_{n\,n-1}b_1 = b_1 e_{n\,n-1}$ that $\beta_1 = \gamma_1 = 0$. Therefore, $b_1 = \beta_2 e_{n1} + \gamma_2 e_{n-2\,n-1} + \delta e_{n\,n-1}$. Obviously, the matrices e_{n1}, $e_{n-2\,n-1}$, and $e_{n\,n-1}$ are pairwise-commutative. Consequently, the set of all nilpotent matrices that commute with a and $e_{n\,n-1}$ constitutes a commutative nilpotent algebra $[a, a^2, \ldots, a^{n-3}, e_{n\,n-1}, e_{n\,1}, e_{n-2\,n-1}]$. From this it follows that

$$N = [a, a^2, \ldots, a^{n-3}, e_{n\,n-1}, e_{n\,1}, e_{n-2\,n-1}]. \tag{3.6}$$

Thus, in case (a, α) we obtain a single algebra (3.6).

Suppose now that case (a, β) obtains. Then, N contains a matrix b_1 of the form (3.5) that does not commute with $e_{n\,n-1}$. Since $b_1 e_{n\,n-1} \neq e_{n\,n-1}b_1$, at least one of the numbers β_1, γ_1 in (3.5) is nonzero.

Lemma 1. *If $e_{n\,n-1} \notin N$, then N contains a basis*

$$a, a^2, \ldots, a^{n-3}, b_1, b_2, \tag{3.7}$$

where

$$b_2 = ab_1 = \beta_1 e_{n\,1} + \gamma_1 e_{n-2\,n-1},$$
$$b_1 = \beta_1(e_{n-1\,1} + e_{n2}) + \beta_2 e_{n1} + \gamma_1(e_{n-3\,n-1} + e_{n-2\,n})$$
$$+ \gamma_2 e_{n-2\,n-1} + \delta e_{n\,n-1}.$$

Proof: N contains an element of the form b_1, where one of the numbers β_1, γ_1 is nonzero. Consequently, $b_2 = ab_1 = \beta_1 e_{n1} + \gamma_1 e_{n-2\,n-1} \in N$. Also, $a, a^2, \ldots, a^{n-3}, b_1, b_2$ are linearly independent. Let us now show that an arbitrary element b of N can be expressed as a linear combination of the elements (3.7). From what has been said (cf. (3.5) and above), we may assume that

$$b = \beta_1'(e_{n-1\,1} + e_{n\,2}) + \beta_2' e_{n\,1} + \gamma_1'(e_{n-3\,n-1} + e_{n-2})$$

$$+ \gamma_2' e_{n-2\,n-1} + \delta e_{n\,n-1}.$$

From the equation $b_1 b = b b_1$, we find

(i) the vectors $(\beta_1', \gamma_1', \delta')$ and $(\beta_1, \gamma_1, \delta)$ are linearly independent, and

$$\text{(ii)} \qquad \gamma_1 \beta_2' + \gamma_2 \beta_1' = \gamma_1' \beta_2 + \gamma_2' \beta_1. \tag{3.8}$$

It follows from (3.8 (i)) that we may assume $\beta_1' = \gamma_1' = \delta' = 0$, that is, $b = \beta_2' e_{n1} + \gamma_2' e_{n-2\,n-1}$. From this and (3.8 (ii)) we get the equation $\gamma_1 \beta_2' = \gamma_2' \beta_1$. Consequently, the vectors (β_2', γ_2') and (β_1, γ_1) are linearly dependent. Therefore, $b = \lambda b_2$ for $\lambda \in P$. This completes the proof of the lemma.

Direct verification shows that the basis (3.7) of the algebra N satisfies the following conditions: $ab_1 = b_2$, $a^2 b_1 = 0$, and

$$b_1{}^2 = \begin{cases} \gamma_1 \beta_1 a^{n-4} + (\gamma_1 \beta_2 + \gamma_2 \beta_1) a^{n-3} + \delta b_2, & n > 5, \\ \gamma_1 \beta_1 a + (\gamma_1 \beta_2 + \gamma_2 \beta_1) a^2 + \delta b_2 + \beta_1 \gamma_1 e_{54}, & n = 5. \end{cases} \tag{3.9}$$

However, for $n = 5$, we have $e_{n\,n-1} = e_{54} \notin N$. Consequently, from (3.9), we have $\beta_1 \gamma_1 = 0$ for $n = 5$.

Lemma 2. *Let c, d_1, and d_2 denote matrices in N such that c^{n-3} and d_2 are linearly independent and $cd_1 = cd_2 = 0$. Then, $c, c^2, \ldots, c^{n-3}, d_1, d_2$ constitute a basis for N.*

Proof: Suppose that

$$\rho_1 c + \rho_2 c^2 + \cdots + \rho_{n-3} c^{n-3} + \rho_{n-2} d_1 + \rho_{n-1} d_2 = 0, \quad \rho_j \in P.$$

$$(3.10)$$

If we multiply (3.10) by c^{n-4}, we obtain $\rho_1 c^{n-3} = 0$ and $\rho_1 = 0$ for $n > 5$. For $n = 5$, we obtain $\rho_1 c^2 + \rho_3 d_2 = 0$ and $\rho_1 = \rho_3 = 0$. Then, from (3.10), we find $\rho_2 c^2 + \rho_4 d_2 = 0$ and $\rho_2 = \rho_4 = 0$ for $n = 5$. Consequently, the lemma is proven for $n = 5$. Suppose now that $n > 5$. Multiplying (3.10) successively by $c^{n-5}, \ldots, c$, we obtain $\rho_2 = \cdots = \rho_{n-4} = \rho_{n-2} = 0$. Then, from (3.10), we find $\rho_{n-3} c^{n-3} + \rho_{n-1} d_2 = 0$ and $\rho_{n-3} = \rho_{n-1} = 0$. This completes the proof of the lemma.

Corollary. *If the matrices c, d_1, and d_2 of the algebra N satisfy the conditions $cd_1 = d_2$, $cd_2 = 0$, and $d_1 d_2 = \alpha c^{n-3} \neq 0$ for $\alpha \in P$, then $c, \ldots, c^{n-3}, d_1, d_2$ constitute a basis for N.*

Proof: If $\rho c^{n-3} + \mu d_2 = 0$, then, by multiplying both sides of the equation by d_1, we obtain $\mu \alpha c^{n-3} = 0$. Therefore, $\mu = \rho = 0$.

Lemma 3. *Let M denote the annihilator of the algebra N. If $\beta_1 \gamma_1 \neq 0$, then $M = [a^{n-3}]$; on the other hand, if $\beta_1 \gamma_1 = 0$, then $M = [a^{n-3}, b_2]$.*

Proof: Suppose that

$$v = \lambda_1 a + \cdots + \lambda_{n-3} a^{n-3} + \mu_1 b_1 + \mu_2 b_2 \in M$$

and $va = 0$. Therefore, $\lambda_1 = \cdots = \lambda_{n-4} = \mu_1 = 0$; that is, $v = \lambda a^{n-3} + \mu b_2$. It follows from (3.9) that $b_1 b_2 = a b_1{}^2 = \gamma_1 \beta_1 a^{n-3}$. Consequently, for $\beta_1 \gamma_1 \neq 0$, we have $v = \lambda a^{n-3}$ and $M = [a^{n-3}]$ and, for $\beta_1 \gamma_1 = 0$, we have $M = [a^{n-3}, b_2]$.

Lemma 4. (i) *If $\beta_1 \gamma_1 \neq 0$, then the signature of N is $(1, n - 2, 1)$.* (ii) *If $\beta_1 \neq 0$ and $\gamma_1 = 0$, then the signature of N is equal to $(1, n - 3, 2)$.* (iii) *If $\beta_1 = 0$ and $\gamma_1 \neq 0$, then the signature of N is $(2, n - 3, 1)$.*

Proof: Suppose that (v, m, μ) is the Kravchuk signature of the algebra N. Then, μv coincides with the dimension of its annihilator M. If $\beta_1 \gamma_1 \neq 0$, then, from Lemma 3 we have $\mu v = 1$ and $\mu = v = 1$. Consequently, (i) is proven. If $\beta_1 \gamma_1 = 0$, then $\mu v = 2$. Consider now the subspace $N P^n$. Obviously, $a P^n = [u_2, u_3, \ldots, u_{n-2}, u_n]$, where $u_1, \ldots, u_n$ constitute the basis for P^n in which the matrices N are written [cf. (3.7)]. If $\beta_1 \neq 0$, then $b_1(u_1) = u_{n-1}$. This means that, for $\beta_1 \neq 0$, we have $N P^n = [u_2, u_3, \ldots, u_{n-1}, u_n]$ and $v = 1$. From this and from the equation $\mu v = 2$, we get (ii). On the other hand, if $\beta_1 = 0$, then $N P^n = [u_2, u_3, \ldots, u_{n-2}, u_n]$. Consequently, $v = 2$ and $\mu = 1$. This completes the proof of the lemma.

Thus, there are three types of algebras N: (i), (ii), and (iii). We note that two algebras belonging to distinct types are not conjugate in P_n since their signatures are distinct. On the other hand, by virtue of Theorem 13 of Chapter 2, for two algebras of a single type to be conjugate in P_n, it is sufficient that they be isomorphic.

Below, we give a classification of the algebras of each of these three types.

(i) The case $\beta_1 \gamma_1 \neq 0$. By virtue of (3.9), this case is possible only for $n > 5$. Obviously, we may take $\beta_1 = 1$.

Then, $b_2 = e_{n\,1} + \gamma_1 e_{n-2\,n-1}$. On the basis of this, we may assume that $\beta_2 = 0$ in the expression (3.7). Thus,

$$b_1 = e_{n-1\,1} + e_{n\,2} + \gamma_1(e_{n-3\,n} + e_{n-2\,n}) + \gamma_2 e_{n-2\,n-1} + \delta e_{n\,n-1}.$$

Then, instead of (3.9), we have

$$\begin{aligned} ab_1 = b_2, \qquad ab_2 = 0 \\ b_1{}^2 = \gamma_1 a^{n-4} + \gamma_2 a^{n-3} + \delta b_2. \end{aligned} \tag{3.11}$$

If instead of b_1 and b_2, we take respectively $b_1/\sqrt{\gamma_1}$ and $b_2/\sqrt{\gamma_1}$, we obtain $\gamma_1 = 1$ in (3.11). Consequently, N possesses the basis $a, a^2, \ldots, a^{n-3}, b_1, b_2$, where $ab_1 = b_2$, $ab_2 = 0$, and $b_1{}^2 = a^{n-4} + \gamma_2 a^{n-3} + \delta b_2$. Now, if we set $a_1 = a + \gamma_2(n-1)^{-1} a^2$, we obtain $b_1{}^2 = a^{n-4} + \delta b_2$. Thus, we obtain a one-parameter family of algebras with basis

$$a_1, a_1{}^2, \ldots, a_1^{n-3}, b_1, b_2, \tag{3.12}$$

where

$$a_1 b_1 = b_2, \quad a_1 b_2 = 0, \text{ and } b_1{}^2 = a_1^{n-4} + \delta b_2.$$

Below, we shall use the function $\psi(x)$ defined on the set of all complex numbers by $\psi(0) = 0$, $\psi(x) = 1$ for $x \neq 0$.

Lemma 5. *If $n > 6$, then $N_\delta \cong N_{\psi(\delta)}$. If $n = 6$ and $\delta \neq \pm 2i$, then $N_\delta \cong N_0$.*

Proof: First, suppose that $n > 6$. For $\delta \neq 0$, we set

$$c = \delta^{2/6-n} a_1, \qquad d_1 = \delta^{4-n/n-6} b_1, \qquad d_2 = \delta^{2-n/n-6} b_2.$$

Then, $cd_1 = d_2$, $cd_2 = 0$, and $d_1{}^2 = c^{n-4} + d_2$. Consequently, $N_\delta \cong N_1$.

Suppose now that $n = 6$ and $\delta^2 + 4 \neq 0$. Let us show that $N_\delta \cong N_0$. We replace the basis (3.12) with a basis

$$c, c^2, c^3, d_1, d_2 \tag{3.13}$$

such that

$$d_1{}^2 = c^2, \qquad cd_1 = d_2, \qquad cd_2 = 0. \tag{3.14}$$

Let us show that $c_1 d_1$ can be chosen so that $c = xa_1 + b_1$ and $d_1 = y_1 a_1 + y_2 b_1$, where $x, y_1, y_2 \in P$. Then,

$$c^2 = (x^2 + 1)a_1{}^2 + (2x + \delta)b_2,$$

$$d_1{}^2 = (y_1{}^2 + y_2{}^2)a_1{}^2 + (2y_1 y_2 + y_2{}^2\delta)b_2,$$

and

$$cd_2 = c^2 d_1 = (x^2 + 1)y_1 a_1{}^3 + (2x + \delta)y_2 a_1{}^3$$

$$= [(x^2 + 1)y_1 + (2x + \delta)y_2]a_1{}^3.$$

From this and from (3.14), we obtain

$$\begin{aligned}
x^2 + 1 &= y_1{}^2 + y_2{}^2, \\
2x + \delta &= 2y_1 y_2 + y_2{}^2\delta, \\
(x^2 + 1)y_1 &+ (2x + \delta)y_2 = 0.
\end{aligned} \tag{3.15}$$

We need to prove the existence of a solution of this system such that the matrices (3.13) are linearly independent. In accordance with the corollary to Lemma 2, for this, it will be sufficient to have c^3 not equal to 0. Since $c^3 = (x^3 + 3x + \delta)a_1{}^3$, we need a solution of the system (3.15) such that

$$x^3 + 3x + \delta \neq 0. \tag{3.16}$$

If we set $y_1 = - z(2x + \delta)$ and $y_2 = z(x^2 + 1)$, then the last of Eqs. (3.15) is satisfied and the first two take the forms

$$f(z) = [(2x + \delta)^2 + (x^2 + 1)^2]z^2 - (x^2 + 1) = 0,$$

$$\varphi(z) = [- 2(2x + \delta) + \delta(x^2 + 1)](x^2 + 1)z^2 - (2x + \delta) = 0.$$

$$(3.17)$$

We note that the coefficients of highest powers of the polynomials $f(z)$ and $\varphi(z)$ do not vanish simultaneously. To see this, suppose, to the contrary, that

$$(2x + \delta)^2 + (x^2 + 1)^2 = 0$$
$$(x^2 + 1)[\delta(x^2 + 1) - 2(2x + \delta)] = 0.$$
$$(3.18)$$

If $x^2 + 1 = 0$, then from (3.18) we obtain $2x + \delta = 0$, where $\delta = \pm 2i$. On the other hand, if $x^2 + 1 \neq 0$, let us rewrite (3.18) as follows:

$$[2x + \delta + i(1 + x^2)][2x + \delta - i(1 + x^2)] = 0$$
$$(x^2 + 1)\delta - 2(2x + \delta) = 0.$$
$$(3.19)$$

The system (3.19) is equivalent to one of the following systems:

$$2x + \delta + i(1 + x^2) = 0, \qquad (x^2 + 1)\delta - 2(2x + \delta) = 0$$
$$(3.20)$$

or

$$2x + \delta - i(1 + x^2) = 0, \qquad (x^2 + 1)\delta - 2(2x + \delta) = 0.$$

Since

$$\begin{vmatrix} 1 & i \\ -2 & \delta \end{vmatrix} = \delta + 2i \neq 0$$

and

$$\begin{vmatrix} 1 & -i \\ -2 & \delta \end{vmatrix} = \delta - 2i \neq 0,$$

it follows from (3.20) that $2x + \delta = 0$ and $x^2 + 1 = 0$, where $\delta = \pm 2i$. This contradicts our assumption.

Let us return to the system (3.17). Since the coefficients of the highest powers of the polynomials $f(z)$ and $\varphi(z)$ do not vanish simultaneously, the system (3.17) has a solution if x is a root of the equation

$$\begin{vmatrix} (2x + \delta)^2 + (x^2 + 1)^2 & x^2 + 1 \\ (x^2 + 1)[\delta(x^2 + 1) - 2(2x + \delta)] & 2x + \delta \end{vmatrix} = 0, \quad (3.21)$$

which we may rewrite

$$(x^3 + 3x + \delta)[\delta x^3 - 6x^2 - 3\delta x - (2 + \delta^2)] = 0.$$

Consider the equation

$$t(x) = \delta x^3 - 6x^2 - 3\delta x - (2 + \delta^2) = 0.$$

We note that the polynomial $t(x)$ has no multiple root when $\delta \neq 0$ and $\delta^2 + 4 \neq 0$. Obviously, in addition, $t(x)$ does not divide $m(x) = x^3 + 3x + \delta$. From this it follows that the polynomial $t(x)$ has a root that is not a root of $m(x)$; that is, Eq. (3.21) has a root satisfying condition (3.16). Thus, the system (3.17) and, consequently, the system (3.15), have a solution satisfying condition (3.16). This completes the proof of the lemma.

Now, let us look at the case $n = 6$ with $\delta^2 + 4 = 0$. It is easy to see that $N_\delta \cong N_{-\delta}$. Consequently, for $\delta^2 + 4 = 0$, we have $N_\delta \cong N_{2i}$.

Lemma 6. (i) *If* $n > 6$, *then* $N_1 \not\cong N_0$. (ii) *If* $n = 6$, *then* $N_0 \not\cong N_{2i}$.

Proof: Suppose that $n > 6$ and $N_1 \cong N_0$. Then, N_1 possesses basis $a_1, \ldots, a_1^{n-3}, b_1, b_2, c, \ldots, c^{n-3}, d_1, d_2$ such that

$$a_1 b_1 = b_2, \qquad a_1 b_2 = 0, \qquad b_1^2 = a_1^{n-4} + b_2,$$
$$cd_1 = d_2, \qquad cd_2 = 0, \qquad d_1^2 = c^{n-4}. \qquad (3.22)$$

Suppose that

$$c = \lambda_1 a_1 + \lambda_2 a_1^2 + \cdots + \lambda_{n-3} a_1^{n-3} + \lambda_{n-2} b_1 + \lambda_{n-1} b_2,$$
$$d_1 = \mu_1 a_1 + \mu_2 a_1^2 + \cdots + \mu_{n-3} a_1^{n-3} + \mu_{n-2} b_1 + \mu_{n-1} b_2.$$

Since $c^2 d_1 = cd_2 = 0$, we have

$$\sum_{i+j+k=e} \lambda_i \lambda_j \mu_k = 0, \qquad e = 3, \ldots, n-4. \qquad (3.23)$$

Furthermore, $c^{n-3} = \lambda_1^{n-3} a_1^{n-3}$. Consequently, $\lambda_1 \neq 0$. From this and (3.23), we see that $\mu_1 = \mu_2 = \cdots = \mu_{n-6} = 0$. Consequently, for d_1, we have

$$d_1^2 = \mu_{n-5}^2 a_1^{2n-10} + \mu_{n-2}^2(a_1^{n-4} + b_2) + 2\mu_{n-2}\mu_{n-1} a_1^{n-3} = c^{n-4}. \qquad (3.24)$$

Since $n > 6$, the expression c^{n-4} in terms of the first basis does not contain b_2. Consequently, $\mu_{n-2} = 0$ and $d_1^2 =$

$\mu_{n-5}^2 a_1^{2n-10}$. If $n > 7$, then $a_1^{2n-10} = 0$. This means that (3.22) is not satisfied. For $n = 7$, we obtain from (3.24)

$$\mu_2^2 a_1^4 = c^3 = \lambda_1^3 a_1^3 + 3(\lambda_1^2\lambda_2 + \lambda_1\lambda_5^2)a_1.$$

Since $\lambda_1 \neq 0$, this last equation is impossible. This proves (i).

To prove (ii), suppose that $n = 6$ and $N_{2i} \cong N_0$. Then, N_{2i} has bases a_1, a_1^2, a_1^3, b_1, b_2 and c, c^2, c^3, d_1, d_2 such that

$$\begin{aligned} a_1b_1 = b_2, \qquad a_1b_2 = 0, \qquad b_1^2 = a_1^2 + 2ib_2, \\ cd_1 = d_2, \qquad cd_2 = 0, \qquad d_1^2 = c^2. \end{aligned} \qquad (3.25)$$

Let us set

$$\begin{aligned} c &= \lambda_1 a_1 + \lambda_2 b_1 + \lambda_3 b_2 + \lambda_4 a_1^2 + \lambda_5 a_1^3, \\ d_1 &= \mu_1 a_1 + \mu_2 b_1 + \mu_3 b_2 + \mu_4 a_1^2 + \mu_5 a_1^3. \end{aligned}$$

Then,

$$\begin{aligned} c^2 &= \lambda_1^2 a_1^2 + \lambda_2^2 b_1^2 + 2(\lambda_1\lambda_2 a_1 b_1 + \lambda_1\lambda_4 a_1^3 + \lambda_2\lambda_3 b_1 b_2) \\ &= (\lambda_1^2 + \lambda_2^2)a_1^2 + 2(\lambda_1\lambda_4 + \lambda_2\lambda_3)a_1^3 + (\lambda_2^2 2i + 2\lambda_1\lambda_2)b_2, \\ cd_2 &= c^2 d_1 = (\lambda_1^2 + \lambda_2^2)\mu_1 a_1^3 + 2(\lambda_2^2 i + \lambda_1\lambda_2)\mu a_1^3. \end{aligned}$$

From (3.25), we obtain

$$\begin{aligned} (\lambda_1^2 + \lambda_2^2)\mu_1 + 2\lambda_2\mu_2(\lambda_1 + i\lambda_2) &= 0, \\ \lambda_1^2 + \lambda_2^2 &= \mu_1^2 + \mu_2^2, \qquad (3.26) \\ \lambda_2(\lambda_1 + i\lambda_2) &= \mu_2(\mu_1 + i\mu_2). \end{aligned}$$

Furthermore,

$$c^3 = [\lambda_1(\lambda_1^2 + \lambda_2^2) + 2\lambda_2^2(\lambda_1 + i\lambda_2)]a_1^3.$$

From this, we get

$$\lambda_1(\lambda_1{}^2 + \lambda_2{}^2) + 2\lambda_2{}^2(\lambda_1 + i\lambda_2) \neq 0. \tag{3.27}$$

Consequently, $\lambda_1 + i\lambda_2 \neq 0$.

It is easy to show that the set of solutions of the system (3.26), (3.27) does not contain a solution such that $\lambda_2 = 0$. Specifically, if $\lambda_2 = 0$, then $\lambda_1 \neq 0$ and $\lambda_1{}^2\mu_1 = 0$. Consequently, $\mu_1 = 0$ and $\mu_2{}^2 = \lambda_1{}^2$. But it follows from the last of the Eqs. (3.26) that $\mu_2 = 0$, so that $\lambda_1 = 0$. But then, (3.27) is not satisfied.

Thus, if the system (3.26), (3.27) has a solution, then, for this solution, $\lambda_2 \neq 0$. Since all equations in the system (3.26) and the left-hand member of inequality (3.27) are homogeneous, it follows on the basis of the preceding considerations that the system (3.26), (3.27) must have a solution with $\lambda_2 = 1$. Setting $\lambda_2 = 1$, we obtain

$$(\lambda_1 - i)\mu_1 + 2\mu_2 = 0. \tag{3.28}$$

$$\lambda_1{}^2 + 1 = \mu_1{}^2 + \mu_2{}^2. \tag{3.29}$$

$$\lambda_1 + i = \mu_2(\mu_1 + i\mu_2). \tag{3.30}$$

$$\lambda_1(\lambda_1{}^2 + 1) + 2(\lambda_1 + i) \neq 0,$$
$$\lambda_1{}^3 + 3\lambda_1 + 2i \neq 0. \tag{3.31}$$

From (3.28), we have $\mu_2 = \frac{1}{2}(i - \lambda_1)\mu_1$. From this and from (3.29),

$$\lambda_1{}^2 + 1 = \mu_1{}^2[1 + \tfrac{1}{4}(\lambda_1 - i)^2]. \tag{3.32}$$

From (3.30),

$$\lambda_1 + i = \mu_1{}^2 \tfrac{1}{2}(i - \lambda_1)[1 + i\tfrac{1}{2}(i - \lambda_1)]$$
$$= \mu_1{}^2[\tfrac{1}{2}(i - \lambda_1) + i\tfrac{1}{4}(i - \lambda_1)^2].$$

Since $\lambda_1 + i \neq 0$, we have $\mu \neq 0$.   Furthermore,

$$\lambda_1{}^2 + 1 = \mu_1{}^2[\tfrac{1}{2}(\lambda_1 - i)(i - \lambda_1) + \tfrac{1}{4}i(i - \lambda_1)^2(\lambda_1 - i)]$$
$$= \mu_1{}^2[-\tfrac{1}{2}(i - \lambda_1)^2 - \tfrac{1}{4}i(i - \lambda_1)^3].$$

Equating this with (3.32) and remembering that $\mu \neq 0$, we obtain

$$1 + \tfrac{1}{4}(i - \lambda_1)^2 + \tfrac{1}{2}(i - \lambda_1)^2 + \tfrac{1}{4}i(i - \lambda_1)^3 = 0,$$
$$4 + 3(i - \lambda_1)^2 + i(i - \lambda_1)^3 = 0,$$
$$2 - 3\lambda_1 i - i\lambda_1{}^3 = 0,$$
$$\lambda_1{}^3 + 3\lambda_1 + 2i = 0.$$

This last contradicts (3.31) and completes the proof of the lemma.

Thus, case (i), when $\beta_1\gamma_1 \neq 0$, is completely studied.   In this case, N is conjugate either with N_0 or N_1 when $n > 6$. When $n = 6$, N is conjugate with either N_0 or N_{2i}.

Before turning to the study of the remaining two cases (ii) and (iii), we note that, on the basis of Theorem 14 of Chapter 2, it will be sufficient to carry out the classification of the algebras in only one of these cases.   Let us consider case (iii).   Then, algebras of type (ii) are obtained by transposing matrices of algebras of type (iii).

(iii) *The case* $\beta_1 = 0$, $\gamma_1 \neq 0$.

Obviously, in this case, we may assume that $\gamma = 1$. In accordance with formulas (3.7), we shall have

$$b_1 = \beta_2 e_{n\,1} + e_{n-3\,n-1} + e_{n-2\,n} + \gamma_2 e_{n-2\,n-1} + \delta e_{n\,n-1},$$
$$b_2 = e_{n-2\,n-1}.$$

Instead of b_1, let us take $b_1 - \gamma_2 b_2$—and then denote *this quantity* by b_1. Thus, N possesses a basis $a, a^2, \ldots, a^{n-3}$, b_1, b_2, where

$$b_1 = \beta_2 e_{n\,1} + e_{n-3\,n-1} + e_{n-2\,n} + \delta e_{n\,n-1}$$

and $b_2 = e_{n-2\,n-1}$. As one can easily verify, $ab_1 = b_2$, $ab_2 = 0$, and

$$b_1{}^2 = \beta_2 e_{n-2\,1} + \delta e_{n-2\,n-1} = \beta_2 a^{n-3} + \delta b_2.$$

Let us set $\beta_2 = \beta$ and $N = N_{\beta,\,\delta}$.

Lemma 7. $N_{\beta,\,\delta} \cong N_{\psi(\beta),\,\psi(\delta)}$ *for $n > 5$.*

Proof: The following four cases are possible: (1) $\beta = 0$, $\delta \neq 0$; (2) $\beta \neq 0$, $\delta = 0$; (3) $\beta\delta \neq 0$; (4) $\beta = \delta = 0$.

(1) Suppose that $\beta = 0$ but $\delta \neq 0$. In the algebra $N_{0,\,\delta}$, let us choose the basis $c, \ldots, c^{n-3}, d_1, d_2$, where $c = a$, $d_1 = \delta^{-1} b_1$, and $d_2 = \delta^{-1} b_2$. Then, $cd_1 = a\delta^{-1} b_1 = \delta^{-1} b_2 = d_2$ and $d_1 = \delta^{-2} b_1{}^2 = \delta^{-1} b_2 = d_2$. Consequently, $N_{0,\,\delta} \cong N_{0,\,1}$.

(2) Suppose that $\beta \neq 0$ but $\delta = 0$. Let us choose in $N_{\beta,\,0}$ a basis $c, \ldots, c^{n-3}, d_1, d_2$, where $c = \lambda a$, $d_1 = b_1$, and $d_2 = \lambda b_2$, where in turn $\lambda^{n-3} = \beta$. Then, $cd_1 = \lambda ab_1 = \lambda b_2 = d_2$ and $d_1{}^2 = b_1{}^2 = \beta a^{n-3} = c^{n-3}$. Consequently, $N_{\beta,\,0} \cong N_{1,\,0}$.

(3) Suppose that β and δ are both nonzero. Let us choose numbers μ and λ in the field P such that $\mu^{n-5} = \beta\delta^{3-n}$ and $\lambda = \mu\delta$. Let us construct a basis for $N_{\beta,\,\delta}$ of the form $c, \ldots, c^{n-3}, d_1, d_2$, where $c = \lambda a$, $d_1 = \mu b_1$, and $d_2 = \lambda\mu b_2$. Then,

$$cd_1 = \lambda\mu ab_1 = \lambda\mu b_2 = d_2$$

and

$$d_1{}^2 = \mu^2 b_1{}^2 = (\beta a^{n-3} + \delta b_2)\mu^2 = \mu^2 \beta a^{n-3} + \mu^2 \delta b_2.$$

On the other hand,

$$c^{n-3} = \lambda^{n-3} a^{n-3}$$

and

$$\lambda^{n-3} = \mu^{n-3}\delta^{n-3} = \mu^2 \mu^{n-5}\delta^{n-3} = \mu^2 \beta.$$

Furthermore,

$$\mu^2 \delta b_2 = \lambda \mu b_2 = d_2.$$

This means that $d_1{}^2 = c^{n-3} + d_2$. Consequently, $N_{\beta, \delta} \cong N_{1,1}$.

(4) If $\beta = \delta = 0$, then the isomorphism $N_{\beta, \delta} \cong N_{\psi(\beta), \psi(\delta)}$ is trivial.

Thus, in all cases, $N_{\beta, \delta} \cong N_{\psi(\beta), \psi(\delta)}$ and the lemma is proven.

Lemma 8. *Suppose that $n = 5$. If $4\beta + \delta^2 \neq 0$, then $N_{\beta, \delta} \cong N_{\psi(\beta), \psi(\delta)}$. On the other hand, if $4\beta + \delta^2 = 0$, then $N_{\beta, \delta} \cong N_{0,0}$.*

Proof: As is clear from the proof of Lemma 7, we need only consider the case $\beta\delta \neq 0$. Thus, suppose that $n = 5$ and $\beta\delta \neq 0$. It is easy to see that $N_{\beta, \delta} \cong N_{1, \delta_1}$, where $\delta_1 = \delta/\sqrt{\beta}$. Specifically, let us choose in $N_{\beta, \delta}$ a basis c, c^2, d_1, d_2, where

$$c = a, \qquad \sqrt{\beta}d_1 = b_1, \qquad \sqrt{\beta}d_2 = b_2.$$

Then,

$$cd_1 = d_2 \qquad \text{and} \qquad d_1^2 = \beta^{-1}b_1^2 = \beta^{-1}(\beta a^2 + \delta b_2)$$

$$= c^2 + \frac{\delta}{\sqrt{\beta}} d_2.$$

Consequently, $N_{\beta, \delta} \cong N_{1, \delta_1}$, where $\delta_1 \sqrt{\beta} = \delta$.

Let us now look at the algebra N_{1, δ_1}, where $\delta_1 \neq 0$. The lemma will be proven if we can establish that $N_{1, \delta_1} \cong N_{1,1}$ for $\delta_1^2 + 4 \neq 0$ and $N_{1, \delta_1} \cong N_{0,0}$ for $\delta_1^2 + 4 = 0$.

Suppose that $\delta_1^2 + 4 \neq 0$. Let us try to find a basis c, c^2, d_1, d_2 of the algebra N_{1, δ_1} such that

$$cd_1 = d_2, \qquad cd_2 = 0, \qquad d_1^2 = c^2 + d_2. \qquad (3.33)$$

Let us seek a basis satisfying conditions (3.33) of the form $c = x_1 a + x_2 b_1$, $d_1 = b_1$, $d_2 = cd_1$, where a, a^2, b_1, b_2 constitute a basis for N_{1, δ_1} such that $ab_1 = b_2$, $ab_2 = 0$, and $b_1^2 = a^2 + \delta_1 b_2$. Then,

$$c^2 = x_1^2 a^2 + x_2^2 b_1^2 + 2x_1 x_2 ab_1$$

$$= x_1^2 a^2 + x_2^2 (a^2 + \delta_1 b_2) + 2x_1 x_2 b_2$$

$$= (x_1^2 + x_2^2)a^2 + (2x_1 x_2 + x_2^2 \delta_1)b_2.$$

$$d_2 = (x_1 a + x_2 b_1)b_1 = x_1 b_2 + x_2(a^2 + \delta_1 b_2)$$

$$= x_2 a^2 + (x_1 + \delta_1 x_2)b_2.$$

In accordance with Lemma 2, the matrices c, c^2, d_1, and d_2 constitute a basis for N_{1, δ_1} if and only if c^2 and d^2 are linearly independent. Consequently,

$$\begin{vmatrix} x_1^2 + x_2^2 & 2x_1 x_2 + x_2^2 \delta_1 \\ x_2 & x_1 + \delta_1 x_2 \end{vmatrix} \neq 0. \qquad (3.34)$$

Condition (3.34) can be rewritten in the form

$$x_1 \neq 0, \qquad \begin{vmatrix} x_1 & x_2 \\ x_2 & x_1 + \delta_1 x_2 \end{vmatrix} \neq 0. \qquad (3.35)$$

Now, we need to show that x_1 and x_2 can be chosen in such a way that $d_1{}^2 = c^2 + d_2$. From the equation $d_1{}^2 = c^2 + d_2$, we find

$$
\begin{aligned}
a^2 + \delta_1 b_2 &= (x_1{}^2 + x_2{}^2)a^2 + (2x_1 x_2 + x_2{}^2 \delta_1)b_2 \\
&\quad + x_2 a^2 + (x_1 + \delta_1 x_2)b_2 \\
&= (x_1{}^2 + x_2{}^2 + x_2)a^2 \\
&\quad + (2x_1 x_2 + x_2{}^2 \delta_1 + x_1 + \delta_1 x_2)b_2.
\end{aligned}
$$

From this, we get

$$x_1{}^2 + x_2{}^2 + x_2 = 1, \qquad 2x_1 x_2 + x_2{}^2 \delta_1 + x_1 + \delta_1 x_2 = \delta_1$$

or

$$
\begin{aligned}
x_1{}^2 &= 1 - x_2 - x_2{}^2 \\
x_1(2x_2 + 1) &= \delta_1(1 - x_2 - x_2{}^2).
\end{aligned}
\qquad (3.36)
$$

Let us show that the system (3.36) has a solution satisfying condition (3.35). Let $x_2 = k_2$ denote a root of the equation

$$\delta_1{}^2(1 - x_2 - x_2{}^2) = (2x_2 + 1)^2. \qquad (3.37)$$

(This equation can be rewritten in the form $(4 + \delta_1{}^2)x_2{}^2 + (4 + \delta_1{}^2)x_2 + 1 - \delta_1 = 0$.) Obviously, k_2 does not make either the right or the left side of Eq. (3.37) vanish. Let us set $x_1 = k_1 = (2k_2 + 1)\delta_1^{-1}$. Then, (k_1, k_2) is a solution of the system (3.36). To see this, note that $2k_2 + 1 = \delta_1 k_1$.

From this and Eq. (3.37), we have $\delta_1{}^2(1 - k_2 - k_2{}^2) = \delta_1{}^2 k_1{}^2$. Consequently, (k_1, k_2) is a solution of the first of Eqs. (3.36). Furthermore, $(2k_2 + 1)^2 = (2k_2 + 1)\delta_1 k_1$. Consequently, $\delta_1{}^2(1 - k_2 - k_2{}^2) = (2k_2 + 1)\delta_1 k_1$; that is, (k_1, k_2) is a solution of the second of Eqs. (3.36). The ordered pair (k_1, k_2) also satisfies inequalities (3.35). We have $k_1 \neq 0$ since $2k_2 + 1 \neq 0$. Also,

$$
\begin{vmatrix}
(2k_2 + 1)\delta_1^{-1} & k_2 \\
k_2 & (2k_2 + 1)\delta_1^{-1} + \delta_1 k_2
\end{vmatrix}
$$

$$
= \delta_1^{-2}
\begin{vmatrix}
2k_2 + 1 & k_2\delta_1 \\
k_2\delta_1 & 2k_2 + 1 + \delta_1{}^2 k_2
\end{vmatrix}
$$

$$
= \delta_1^{-2}[(2k_2 + 1)^2 - (k_2{}^2\delta_1{}^2 - \delta_1{}^2 2k_2{}^2 - \delta_1{}^2 k_2)]
$$

$$
= \delta_1^{-2}[(2k_2 + 1)^2 + \delta_1{}^2(k_2{}^2 + k_2)]
$$

$$
= \delta_1^{-1}[\delta_1{}^2(1 - k_2 - k_2{}^2) + \delta_1(k_2{}^2 + k_2)] = 1.
$$

Consequently, $N_{1,\delta_1} \cong N_{1,1}$ for $\delta_1{}^2 + 4 \neq 0$. It now remains to show that $N_{1,\delta_1} \cong N_{0,0}$ for $\delta_1{}^2 + 4 = 0$. Suppose that $\delta_1{}^2 + 4 = 0$. Let us take as the basis for N_{1,δ_1} matrices c, c^2, d_1, d_2, where $c = a$, $d_1 = b_1 - \frac{1}{2}\delta_1 a$, and $d_2 = b_2 - \frac{1}{2}\delta_1 a^2$. Then, $cd_1 = d_2$, $cd_2 = 0$, and

$$
d_1{}^2 = b_1{}^2 - \delta_1 a b_1 - a^2 = a^2 + \delta_1 b_1 - \delta_1 b_2 - a^2 = 0.
$$

Therefore, $N_{1,\delta_1} \cong N_{0,0}$. This completes the proof of the lemma.

Lemma 9. $N_{1,1} \cong N_{0,1}$.

Proof: Suppose first that $n > 5$. Let us show that $N_{1,1}$ contains a basis $c, \ldots, c^{n-3}, d_1, d_2$ such that $cd_1 = d_2$,

$cd_2 = 0$, and $d_1{}^2 = d_2$. Let us set $c = a$, $d_1 = a^{n-4} + b_1$, and $d_2 = a^{n-3} + b_2$. Then, $cd_1 = d_2$, $cd_2 = 0$, and $d_1{}^2 = b_1{}^2 = a^{n-3} + b_2 = d_2$. For $n > 5$, the lemma is proven.

For $n = 5$, let us set

$$c = \sqrt{5}a$$
$$d_1 = \tfrac{1}{2}(-1 + \sqrt{5})a + b_1,$$
$$d_2 = \tfrac{1}{2}(5 - \sqrt{5})a^2 + \sqrt{5}b_2.$$

Then, c, c^2, d_1, d_2 is a basis for $N_{1,1}$ such that $cd_1 = d_2$, $cd_2 = 0$, and

$$
\begin{aligned}
d_1{}^2 &= (\tfrac{1}{2}(-1 + \sqrt{5}))^2 a^2 + b_1{}^2 + (\sqrt{5} - 1)ab_1 \\
&= \tfrac{1}{4}(1 + 5 - 2\sqrt{5})a^2 + a^2 + b^2 + (\sqrt{5} - 1)b_2 \\
&= \tfrac{1}{2}(5 - \sqrt{5})a^2 + \sqrt{5}b_2 = d_2.
\end{aligned}
$$

Consequently, $N_{1,1} \cong N_{0,1}$ for $n = 5$ also. This completes the proof of the lemma.

Lemma 10. *If $n > 5$, then $N_{0,1} \cong N_{1,0}$ and $N_{0,1} \cong N_{0,0}$. If $n = 5$, then $N_{0,1} \not\cong N_{0,0}$ but $N_{0,1} \cong N_{1,0}$.*

Proof: Suppose first that $n > 5$ and that either $N_{0,1} \cong N_{1,0}$ or $N_{0,1} \cong N_{0,0}$. Then, $N_{0,1}$ contains a basis $c, \ldots, c^{n-3}, d_1, d_2$ such that $cd_1 = d_2$, $cd_2 = 0$, and $d_1{}^2 = c^{n-3}$ if $N_{0,1} \cong N_{1,0}$ but $d_1{}^2 = 0$ if $N_{0,1} \cong N_{0,0}$. Let us seek c, d_1, and d_2 in the forms

$$
\begin{aligned}
c &= x_1 a + \cdots + x_{n-3}a^{n-3} + x_{n-2}b_1 + x_{n-1}b_2, \\
d_1 &= y_1 a + \cdots + y_{n-3}a^{n-3} + y_{n-2}b_1 + y_{n-1}b_2, \\
d_2 &= cd_1 = x_1 y_1 a^2 + \cdots + \sum_{i+j=n-3} x_i y_j a^{n-3} \\
&\quad + (x_1 y_{n-2} + x_{n-2} y_1 + x_{n-2} y_{n-2})b_2.
\end{aligned}
\tag{3.38}
$$

From the condition

$$cd_2 = 0, \tag{3.39}$$

we then find

$$x_1^2 y_1 = \cdots = \sum_{i+j+k=n-3} x_i x_j y_k = 0. \tag{3.40}$$

Since $c^{n-3} = x^{n-3} a^{n-3} \neq 0$, it follows that $x_1 \neq 0$ and the equations $y_1 = \cdots = y_{n-5} = 0$ then follow from (3.40). Consequently, $d_2 = x_1 y_{n-4} a^{n-3} + (x_1 + x_{n-2}) y_{n-2} b_2$. Since c^{n-3} and d_2 are linearly independent, we have $y_{n-2} \neq 0$ and

$$d_1^2 = (y_{n-4} a^{n-4} + y_{n-3} a^{n-3} + y_{n-2} b_1 + y_{n-1} b_2)^2$$
$$= y_{n-2}^2 b_1^2 = y_{n-2}^2 b_2,$$

that is, $d_1^2 \neq c^{n-3}$ and $d_1^2 \neq 0$. This proves the lemma for $n > 5$.

Suppose now that $n = 5$. Let us show first that $N_{0,1} \not\cong N_{0,0}$. If $N_{0,1} \cong N_{0,0}$, then $N_{0,1}$ contains a basis c, c^2, d_1, d_2 such that $cd_1 = d_2$ and $d_1^2 = 0$. Let us seek matrices c and d_1 of the form (3.38). Then,

$$d_2 = x_1 y_1 a^2 + (x_1 y_3 + x_3 y_1 + x_3 y_3 b_2)$$

and

$$c^2 = x_1^2 a^2 + (2x_1 x_3 + x_3^2) b_2.$$

Since c^2 and d_2 are linearly independent, we have

$$\begin{vmatrix} x_1^2 & 2x_1 x_3 + x_3^2 \\ x_1 y_1 & x_1 y_3 + x_3 y_1 + x_3 y_3 \end{vmatrix} \neq 0. \tag{3.41}$$

Furthermore, $d_1{}^2 = y_1{}^2 a^2 + (2y_1 y_2 + y_3{}^2)b_2$. Since $d_1{}^2 = 0$, we have $y_1 = y_3 = 0$. This last contradicts (3.41). Consequently, $N_{0,1} \not\cong N_{0,0}$. Finally, let us show that $N_{0,1} \cong N_{1,0}$ for $n = 5$. Let us set $c = a$, $d_1 = a - 2b_1$, and $d_2 = cd_1 = a^2 - 2b_2$. Then,

$$d_1{}^2 = a^2 - 4ab_1 + 4b_1{}^2 = a^2 - 4b_2 + 4b_2 = a^2 = c^2.$$

Consequently, $N_{0,1} \cong N_{1,0}$. This completes the proof of the lemma.

Lemma 11. $N_{0,1} \not\cong N_{0,0}$ *for* $n > 5$.

Proof: Suppose that $N_{1,0} \cong N_{0,0}$. Then, $N_{1,0}$ contains a basis $c, \ldots, c^{n-3}, d_1, d_2$ such that $cd_1 = d_2$, $cd_2 = 0$, and $d_1{}^2 = 0$. Let us seek c and d_1 in the form (3.38). Then, from condition (3.39), we shall obtain Eqs. (3.40). Furthermore,

$$d_2 = (x_1 y_{n-4} + x_{n-2} y_{n-2})a^{n-3} + x_1 y_{n-2} b_2.$$

Consequently, $y_{n-2} \neq 0$. Therefore, $d_1{}^2 = y_{n-2}a^{n-3} \neq 0$. This completes the proof of the lemma.

Thus, consideration of case (iii), when $\beta_1 = 0$ but $\gamma_1 \neq 0$, leads to the following result:

For $n > 5$, the algebra $N_{\beta, \delta}$ is conjugate with one of the three pairwise-nonisomorphic algebras: $N_{1,0}, N_{0,1}, N_{0,0}$. For $n = 5$, the algebra $N_{\beta, \delta}$ is conjugate with one of the two nonisomorphic algebras: $N_{1,0}, N_{0,0}$.

As was noted above, type (ii) algebras are easily constructed when $\beta_1 \neq 0$ but $\gamma_1 = 0$.

Lemma 12. For $n > 5$, a type (ii) algebra is conjugate with one of the three pairwise-nonisomorphic algebras: $N'_{1,0}$,

$N'_{0,1}$, $N'_{0,0}$. *For* $n = 5$, *a type* (ii) *algebra is conjugate with one of the two nonisomorphic algebras:* $N'_{1,0}$, $N'_{0,0}$.

Here, N' is an algebra obtained from N by transposing all its matrices.

Thus, our study of case (a) is complete.

Let us turn now to case (b) when N contains no matrix a of rank $n - 2$ such that $a^{n-3} \neq 0$.

Suppose that $a \in N$ and $a^{n-3} \neq 0$. Then, we may assume that $a = e_{21} + e_{32} + \cdots + e_{n-2\,n-3}$. If $b \in N$, then $ab = ba$ and b is nilpotent. Consequently, b is of the form

$$b = \begin{bmatrix} 0 & 0 & \ldots & 0 & 0 & 0 & 0 \\ \alpha_1 & 0 & \ldots & 0 & 0 & 0 & 0 \\ \multicolumn{7}{c}{\cdots\cdots\cdots\cdots\cdots\cdots\cdots\cdots\cdots\cdots} \\ \alpha_{n-3} & \alpha_{n-4} & \ldots & \alpha_1 & 0 & \gamma & \delta \\ \alpha & 0 & \ldots & 0 & 0 & \lambda & \mu \\ \beta & 0 & \ldots & 0 & 0 & \nu & \varepsilon \end{bmatrix}. \qquad (3.42)$$

Since b is nilpotent, the matrix

$$\begin{bmatrix} \lambda & \mu \\ \nu & \varepsilon \end{bmatrix}$$

in formula (3.42) is also nilpotent.

Lemma 13. *For an arbitrary matrix* $b \in N$, *we have* $\lambda = \mu = \nu = \varepsilon = 0$.

Proof: Suppose that N contains a matrix b of the form (3.40) such that

$$\begin{bmatrix} \lambda & \mu \\ \nu & \varepsilon \end{bmatrix} \neq 0.$$

Then, N contains a matrix of rank $n-2$, for example,

$$a_1 = b - \alpha_1 a - \alpha_2 a^2 - \cdots - \alpha_{n-3} a^{n-3} + a.$$

Consider $a + \tau a_1$ for $\tau \in P$. If $\tau \neq 0$, then the rank of $a + \tau a_1$ is equal to $n-2$. On the other hand, the equation $(a + \tau a_1)^{n-3} = 0$ is not an identity in τ since $(a + \tau a_1)^{n-3} = a^{n-3} \neq 0$ for $\tau = 0$. Consequently, there exist infinitely many values of $\tau \in P$ such that $(a + \tau a_1)^{n-3} \neq 0$. Thus, N contains a matrix a_2 of rank $n-2$ such that $a_2^{n-3} \neq 0$. This last contradicts the assumption and proves the lemma.

Therefore, for a basis for N, we can choose the matrices

$$a, a^2, a^{n-3}, b_1, b_2, \ldots, \tag{3.43}$$

where the matrices $b_1, b_2, \ldots$ are linear combinations of the matrices $e_{n-1\,1}, e_{n\,1}, e_{n-2\,n-1}, e_{n-2\,n}$. It is easy to show that there are exactly two matrices b_i in (3.43). Specifically,

$$b_1 = \alpha e_{n-1\,1} + \beta e_{n\,1} + \gamma e_{n-2\,n-1} + \delta e_{n-2\,n} \tag{3.44}$$

belongs to N and $b_1 \neq 0$. Also,

$$b = x_1 e_{n-1\,1} + x_2 e_{n\,1} + x_3 e_{n-2\,n-1} + x_4 e_{n-2\,n}$$

belongs to N. Then, from the condition $bb_1 = b_1 b$, we find that $\alpha x_3 + \beta x_4 - \gamma x_1 - \delta x_2 = 0$. Consequently, every maximal subspace of pairwise-commutative matrices of the form (3.44) is of dimension 2. Thus, N possesses a basis of the form

$$a, a^2, \ldots, a^{n-3}, b_1, b_2, \tag{3.45}$$

where

$$b_1 = \alpha e_{n-1\,1} + \beta e_{n\,1} + \gamma e_{n-2\,n-1} + \delta e_{n-2\,n},$$

$$b_2 = \alpha_1 e_{n-1\,1} + \beta_1 e_{n\,1} + \gamma_1 e_{n-2\,n-1} + \delta_1 e_{n-2\,n}, \quad (3.46)$$

$$\alpha\gamma_1 + \beta\delta_1 = \gamma\alpha_1 + \delta\beta_1$$

and the vectors $(\alpha, \beta, \gamma, \delta)$ and $(\alpha_1, \beta_1, \gamma_1, \delta_1)$ are linearly independent. The algebra N, whose basis (3.45) satisfies conditions (3.46) is maximal among all the nilpotent commutative subalgebras of P_n that do not contain a matrix a of rank $n - 2$ satisfying condition (3.4).

Lemma 14. *An algebra N with basis (3.45) is maximal among all the commutative nilpotent subalgebras of P_n if and only if not one of the three matrices*

$$e_{n-1\,n}, \qquad e_{n\,n-1}, \qquad e_{n-1\,n-1} - e_{n\,n} + \mu e_{n-1\,n} - \mu^{-1} e_{n\,n-1},$$

$$\mu \in P, \qquad \mu \neq 0 \qquad\qquad (3.47)$$

commutes with both b_1 and b_2.

Proof: Suppose that N is nonmaximal. Then, $P_n \backslash N$ contains a nilpotent matrix b that commutes with a, b_1, and b_2. On the basis of (3.42), we may assume that b is of the form

$$b = \alpha_2 e_{n-1\,1} + \beta_2 e_{n\,1} + \gamma_2 e_{n-2\,n-1} + \delta_2 e_{n-2\,n}$$

$$+ \lambda e_{n-1\,n-1} + \mu e_{n-1\,n} + \nu e_{n\,n-1} - \lambda e_{n\,n},$$

where

$$\begin{bmatrix} \lambda & \mu \\ \nu & -\lambda \end{bmatrix}$$

is a nilpotent matrix. The conditions $bb_1 = b_1 b$ and $bb_2 = b_2 b$ imply that the matrix

$$b_3 = \alpha_2 e_{n-1\,1} + \beta_2 e_{n\,1} + \gamma_2 e_{n-1\,n-1} + \delta_2 e_{n-2\,n}$$

commutes with b_1 and b_2. Consequently, b_3 is a linear combination of b_1 and b_2. This means that $P_n \backslash N$ contains a matrix of the form

$$c = \lambda e_{n-1\,n-1} + \mu e_{n-1\,n} + \nu e_{n\,n-1} - \lambda e_{n\,n}.$$

Since

$$\begin{bmatrix} \lambda & r \\ \nu & -\lambda \end{bmatrix}$$

is nilpotent, we have $\lambda^2 + \mu\nu = 0$. If $\lambda = 0$, then $\mu\nu = 0$. Consequently, either $c = \mu e_{n-1\,n}$ or $c = \nu e_{n\,n-1}$. If $\lambda = 1$, then

$$c = e_{n-1\,n-1} - e_{n\,n} + e_{n-1\,n} - \mu^{-1} e_{n\,n-1}.$$

This completes the proof of the lemma.

Let ρ denote the rank of the matrix

$$\begin{bmatrix} \alpha & \beta \\ \alpha_1 & \beta_1 \end{bmatrix}$$

[cf. (3.46)]. Three cases are possible: $\rho = 2$; $\rho = 0$; and $\rho = 1$.

Suppose that $\rho = 2$. Then, we may assume that $\alpha = \beta_1 = 1$ and $\alpha_1 = \beta = 0$. From condition (3.46), we obtain $\gamma_1 = \delta$. Consequently,

$$b_1 = e_{n-1\,1} + \gamma e_{n-2\,n-1} + \delta e_{n-2\,n}$$

and

$$b_2 = e_{n\,1} + \delta e_{n-2\,n-1} + \delta_1 e_{n-2\,n}.$$

Not one of the matrices (3.47) commutes simultaneously with b_1 and b_2. Consequently, for arbitrary γ, δ, and δ_1, the algebra N is, on the basis of Lemma 14, a maximal commutative nilpotent subalgebra of P_n. The multiplication table in N has the form

$$ab_1 = ab_2 = 0, \qquad b_1{}^2 = \gamma a^{n-3},$$
$$b_1 b_2 = \delta a^{n-3}, \qquad b_2{}^2 = \delta_1 a^{n-3}.$$

Let us put the symmetric matrix

$$A = \begin{bmatrix} \gamma & \delta \\ \delta & \delta_1 \end{bmatrix}$$

in correspondence with the four products $b_1{}^2$, $b_1 b_2$, $b_2 b_1$, and $b_2{}^2$. If we carry out a nonsingular linear transformation on b_1 and b_2, the matrix A changes like the matrix of a quadratic form (cf. Section 8, Chapter 2). Therefore, we may assume: (i) that $A = 0$; (ii) that $A = [1, 0]$; or (iii) that $A = E_2$.

Lemma 15. *In case* (i), *N has the signature* $(1, n - 4, 3)$. *In case* (ii), *N has the signature* $(1, n - 3, 2)$. *In case* (iii), *N has the signature* $(1, n - 3, 1)$.

Proof: $aP^n = [u_2, \ldots, u_{n-2}]$, where $u_1, \ldots, u_n$ is that basis of the space P^n in which the matrices (3.45) are written. Furthermore, $b_1(u_1) = u_{n-1}$ and $b_2(u_1) = u_n$. Consequently, $NP^n = [u_2, \ldots, u_n]$, the dimension $NP^n : P = n - 1$, where 1 is the first number in the signature of the algebra N. It only remains to calculate the dimension of the annihilator

M of the algebra N. In case (i), $M = [a^{n-3}, b_1, b_2]$, $\mu\nu = 3$, and $(1, n - 4, 3)$ is the signature of N. In case (ii), $M = [a^{n-3}, b_2]$, $\mu\nu = 2$, and $(1, n - 3, 2)$ is the signature of N. In case (iii), $M = [a^{n-3}]$, $\mu\nu = 1$, and $(1, n - 2, 1)$ is the signature of N. This completes the proof of the lemma. Now, we obviously have

Lemma 16. *If* $\rho = 2$, *then* N *is conjugate in* P_n *with one of the three pairwise-nonconjugate subalgebras*

$$N_1 = [a, \ldots, a^{n-3}, e_{n-1\,1}, e_{n\,1}], \tag{3.48}$$

$$N_2 = [a, \ldots, a^{n-3}, e_{n-1\,1} + e_{n-2\,n-1}, e_{n\,1}], \tag{3.49}$$

$$N_3 = [a, \ldots, a^{n-3}, e_{n-1\,1} + e_{n-2\,n-1}, e_{n\,1} + e_{n-2\,n}]. \tag{3.50}$$

Suppose that $\rho = 0$. Then,

$$b_1 = \gamma e_{n-2\,n-1} + \delta e_{n-2\,n},$$

and

$$b_2 = \gamma_1 e_{n-2\,n-1} + \delta_1 e_{n-2\,n}.$$

The matrices b_1 and b_2 are linearly independent. Therefore, we may assume that $b_1 = e_{n-2\,n-1}$ and $b_2 = e_{n-2\,n}$. Consequently,

$$N = [a, \ldots, a^{n-3}, e_{n-2\,n-1}, e_{n-2\,n}]. \tag{3.51}$$

Not one of the matrices (3.47) commutes with both $e_{n-2\,n-1}$ and $e_{n-2\,n}$. Therefore, in accordance with Lemma 15, the algebra (3.51) is a maximal commutative nilpotent subalgebra of P_n. Obviously, the annihilator M of the algebra N consists of matrices of the form

$$\alpha a^{n-3} + \beta e_{n-2\,n-1} + \gamma e_{n-2\,n}$$

and, consequently, $M : P = 3$. Furthermore, $NP^n : P = 3$. Therefore $(3, n - 4, 1)$ is the signature of the algebra (3.51).

Suppose that $\rho = 1$. Then. we may assume that $\alpha_1 = \beta_1 = 0$. From conditions (3.46), it then follows that

$$\begin{vmatrix} \alpha & \beta \\ -\delta_1 & \gamma_1 \end{vmatrix} = 0.$$

Therefore, we can take

$$b_1 = \alpha e_{n-1\,1} + \beta e_{n\,1} + \gamma e_{n-2\,n-1} + \delta e_{n-2\,n},$$
$$b_2 = \beta e_{n-2\,n-1} - \alpha e_{n-2\,n}. \tag{3.52}$$

Direct calculations show that a necessary and sufficient condition for each of the matrices (3.47) to fail to commute with at least one of $b_1 b_2$ is that

$$\begin{vmatrix} \alpha & -\beta \\ \delta & \gamma \end{vmatrix} \neq 0. \tag{3.53}$$

Thus, N is maximal among the commutative nilpotent subalgebras of P_n if and only if inequality (3.53) holds.

On the basis of (3.52), N has the multiplication table

$$ab_1 = ab_2 = 0, \qquad b_1{}^2 = (\alpha\gamma + \beta\delta)a^{n-3}, \qquad b_1 b_2 = b_2{}^2 = 0. \tag{3.54}$$

It follows from (3.53) and (3.54) that the dimension of the annihilator of the algebra N is 2. But then $(2, n - 3, 1)$ is the signature of N. It now follows from the preceding considerations that N is conjugate with the algebra N_2' where N_2 is the algebra (3.49) and the prime indicates that the matrices have been transposed.

We note also that an algebra N of the form (3.51) is conjugate with N_1', where N_1 is the algebra (3.48). Thus we have completed the study of case (b).

All these results can be consolidated in

Theorem 2. *The set of all maximal commutative nilpotent subalgebras of Class* $n - 2$ *of the algebra* P_n *can be partitioned into* 14 (10) *classes of conjugate elements for* $n > 5$ *(for* $n = 5$*). For a system of representatives of these classes, we can choose the following algebras:*

For $n > 6$:

$$N_1 = [a, \ldots, a^{n-3}, e_{n\,n-1}, e_{n\,1}, e_{n-2\,n-1}],$$

where

$$a = e_{21} + e_{32} + \cdots + e_{n-2\,n-3}; \qquad (3.55)$$

$$N_2 = [a, a^2, \ldots, a^{n-3}, e_{n-1\,1} + e_{n\,2} + e_{n-3\,n-1}$$
$$+ e_{n-2\,n}, e_{n\,1} + e_{n-2\,n-1}],$$

where

$$a = e_{21} + e_{32} + \cdots + e_{n-2\,n-3} + e_{n\,n-1}; \qquad (3.56)$$

$$N_3 = [a, a^2, \ldots, a^{n-3}, e_{n-1\,1} + e_{n\,2} + e_{n-3\,n-1}$$
$$+ e_{n-2\,n} + e_{n\,n-1}, e_{n\,1} + e_{n-2\,n-1}],$$

where a *is the martix* (3.56).

$$N_4 = [a, a^2, \ldots, a^{n-3}, e_{n\,1} + e_{n-3\,n-1} + e_{n-2\,n}, e_{n-2\,n-1}],$$

where a *is the matrix* (3.56).

$$N_5 = [a, a^2, \ldots, a^{n-3}, e_{n-3\,n-1} + e_{n-2\,n} + e_{n\,n-1}, e_{n-2\,n-1}],$$

where a is the matrix (3.56).

$$N_6 = [a, a^2, \ldots, a^{n-3}, e_{n-3\,n-1} + e_{n-2\,n}, e_{n-2\,n-1}],$$

where a is the matrix (3.56).

$$N_7 = N_4';$$

$$N_8 = N_5';$$

$$N_9 = N_6';$$

$$N_{10} = [a, \ldots, a^{n-3}, e_{n-1\,1}, e_{n\,1}],$$

where a is the matrix (3.55);

$$N_{11} = [a, \ldots, a^{n-3}, e_{n-1\,1} + e_{n-2\,n-1}, e_{n\,1}],$$

where a is the matrix (3.55);

$$N_{12} = N_{10}';$$

$$N_{13} = N_{11}';$$

$$N_{14} = [a, a^2, \ldots, a^{n-3}, e_{n-1\,1} + e_{n-2\,n-1}, e_{n\,1} + e_{n-2\,n}],$$

where a is the matrix (3.55).

For $n = 6$:

$$N_1 = [a, a^2, a^3, e_{65}, e_{61}, e_{45}],$$

where

$$a = e_{21} + e_{32} + e_{43}; \tag{3.57}$$

$$N_2 = [a, a^2, a^3, e_{51} + e_{62} + e_{35} + e_{46}, e_{61} + e_{45}],$$

where

$$a = e_{21} + e_{32} + e_{43} + e_{65}; \tag{3.58}$$

$$N_3 = [a, a^2, a^3, e_{51} + e_{62} + e_{35} + e_{46} + 2ie_{65}, e_{61} + e_{45}],$$

where a *is the matrix* (3.58).

$$N_4 = [a, a^2, a^3, e_{61} + e_{35} + e_{46} + e_{45}],$$

where a *is the matrix* (3.58);

$$N_5 = [a, a^2, a^3, e_{35} + e_{46} + e_{65}, e_{45}],$$

where a *is the matrix* (3.58);

$$N_6 = [a, a^2, a^3, e_{35} + e_{46}, e_{45}],$$

where a *is the matrix* (3.58);

$$N_7 = N_4';$$
$$N_8 = N_5';$$
$$N_9 = N_6';$$
$$N_{10} = [a, a^2, a^3, e_{51}, e_{61}],$$

where a *is the matrix* (3.57);

$$N_{11} = [a, a^2, a^3, e_{51} + e_{45}, e_{61}],$$

where a *is the matrix* (3.57);

$$N_{12} = N_{10}';$$
$$N_{13} = N_{11}';$$
$$N_{14} = [a, a^2, a^3, e_{51} + e_{45}, e_{61} + e_{46}],$$

where a *is the matrix* (3.57).

For $n = 5$:

$$N_1 = [a, a^2, e_{54}, e_{51}, e_{34}], \qquad (3.59)$$

where

$$a = e_{21} + e_{32};$$

$$N_2 = [a, a^2, e_{51} + e_{24} + e_{35}, e_{34}],$$

where

$$a = e_{21} + e_{32} + e_{54};$$

$$N_3 = [a, a^2, e_{24} + e_{35}, e_{34}],$$

(3.60)

where a is the matrix (3.60);

$$N_4 = N_2';$$
$$N_5 = N_3';$$
$$N_6 = [a, a^2, e_{41}, e_{51}],$$

where a is the matrix (3.59);

$$N_7 = [a, a^2, e_{41} + e_{34}, e_{51}],$$

where a is the matrix (3.59);

$$N_8 = N_6';$$
$$N_9 = N_7';$$
$$N_{10} = [a, a^2, e_{41} + e_{34}, e_{51} + e_{35}],$$

where a is the matrix (3.59).

Here, N' is the algebra consisting of the transposes of the matrices of N and $\alpha_j, \beta, \gamma, \delta \in P$.

3. Commutative Matrix Algebras of Low Order

Let P_n denote the algebra of all $n \times n$ matrices over the field P of complex numbers. In accordance with Theorem 16 of Chapter 2, for $n > 6$, the algebra P_n has infinitely many pairwise-nonconjugate maximal commutative nilpotent subalgebras. Below, we shall give a complete description of the maximal commutative nilpotent subalgebras of P_n for $n \leqslant 6$. In other words, for $n \leqslant 6$, we shall draw up a table of pairwise-nonconjugate maximal commutative nilpotent subalgebras of P_n such that each maximal nilpotent subalgebra of the algebra P_n is conjugate with one of the algebras of the table. With the aid of obvious constructions from this table we can obtain a table of maximal commutative subalgebras of the algebra P_n for $n \leqslant 6$. In particular, from the considerations that we point out, it follows that, for $n \leqslant 6$, the algebra P_n has only finitely many pairwise-nonconjugate maximal commutative subalgebras. Combining the results of the present section with the results of Section 1 of Chapter 2, we can obtain a description of all maximal commutative subgroups of $GL(n, P)$ where $n \leqslant 6$.

In all that follows, l denotes the class of nilpotency of the algebra in question. As was shown in Section 2 of Chapter 2, for $l = 2, 3, \ldots, n$, the algebra P_n contains maximal commutative nilpotent subalgebras of class l. The maximal commutative nilpotent subalgebras of P_n that are of class l have been described for the case in which $l = 2, n, n - 1$, and $n - 2$ (cf. Sections 3 and 4 of Chapter 2 and Section 1 of Chapter 3). Consequently, for $n \leqslant 6$, the table of algebras can be immediately constructed by use of the results discussed except in the case of $n = 6$, $l = 3$. To study this one case, we shall use the table of commutative nilpotent algebras of dimension 5 of Class 3 that was constructed in Section 9 of Chapter 2.

Let us turn now to the construction of the table.

I. $n = 2$. Up to conjugacy, P_2 contains a unique maximal nilpotent commutative algebra:

$$N_1 = [e_{21}].$$

II. $n = 3$. Then, two cases are possible: $l = 2$; $l = 3$. For $l = 2$,

$$N_1 = [e_{21}, e_{31}],$$
$$N_2 = [e_{31}, e_{32}];$$

for $l = 3$,

$$N_3 = [e_{21} + e_{32}, e_{31}].$$

III. $n = 4$. Three cases are possible: $l = 2$; $l = 3$; $l = 4$. For $l = 2$,

$$N_1 = [e_{21}, e_{31}, e_{41}],$$
$$N_2 = [e_{31}, e_{32}, e_{41}, e_{42}],$$
$$N_3 = [e_{41}, e_{42}, e_{43}];$$

for $l = 3$,

$$N_4 = [e_{21} + e_{32}, e_{31}, e_{41}],$$
$$N_5 = [e_{21} + e_{32}, e_{31}, e_{34}],$$
$$N_6 = [e_{21} + e_{32}, e_{31} + e_{34}];$$

for $l = 4$,

$$N_7 = [e_{21} + e_{32} + e_{43}, e_{31} + e_{42}, e_{41}].$$

IV. $n = 5$. Four cases are possible: $l = 2$; $l = 3$; $l = 4$; $l = 5$.

For $l = 2$,

$$N_1 = [e_{21}, e_{31}, e_{41}, e_{51}],$$
$$N_2 = [e_{31}, e_{32}, e_{41}, e_{42}, e_{51}, e_{52}],$$
$$N_3 = [e_{41}, e_{42}, e_{43}, e_{51}, e_{52}, e_{53}],$$
$$N_4 = [e_{51}, e_{52}, e_{53}, e_{54}].$$

For $l = 3$, we obtain, in accordance with Theorem 2,

$$N_5 = [a, a^2, e_{54}, e_{51}, e_{34}],$$

where

$$a = e_{21} + e_{32}; \tag{3.61}$$

$$N_6 = [a, a^2, e_{51} + e_{24} + e_{35}, e_{34}],$$

where

$$a = e_{21} + e_{32} + e_{54}, \tag{3.62}$$

$$N_7 = [a, a^2, e_{24} + e_{35}, e_{34}],$$

where a is a matrix of the form (3.62);

$$N_8 = N_6';$$
$$N_9 = N_7';$$
$$N_{10} = [a, a^2, e_{41}, e_{51}],$$

where a is a matrix of the form (3.61);

$$N_{11} = [a, a^2, e_{41} + e_{34}, e_{51}],$$

where a is a matrix of the form (3.61);

$$N_{12} = N_{10}';$$
$$N_{13} = N_{11}';$$
$$N_{14} = [a, a^2, e_{41} + e_{34}, e_{51} + e_{35}],$$

where a is a matrix of the form (3.61).

Here, N' is obtained from the algebra N by transposing all its matrices.

For $l = 4$,

$$N_{15} = [a, a^2, a^3, e_{51}],$$
$$N_{16} = [a, a^2, a^3, e_{45}],$$
$$N_{17} = [a, a^2, a^3, e_{51} + e_{45}],$$

where $a = e_{21} + e_{32} + e_{43}$.

For $l = 5$,

$$N_{18} = [a, a^2, a^3, a^4],$$

where $a = e_{21} + e_{23} + e_{32} + e_{43} + e_{54}$.

V. $n = 6$. Five cases are possible: $l = 3$; $l = 2$; $l = 4$; $l = 5$; $l = 6$.

For $l = 3$, let (ν, m, μ) denote the Kravchuk signature of the algebra N. Then, $\nu + m + \mu = 6$, where ν, m, and μ are all positive. Here, five subcases are possible:

(a) $\nu = 1$;   (b) $\nu = 2, m = 1, \mu = 3$;   (c) $\nu = m = \mu = 2$;

(d) $\mu = 1, \nu \neq 1$;   (e) $\nu = 3, m = 1, \mu = 2$.

Subcase (a) $\nu = 1$. According to Theorem 13 of Chapter 2, algebras with signature $(1, m, \mu)$ are characterized by the fact that they and only they are conjugate with their regular representation. Consequently, in the present case, it will be sufficient to construct regular representations of all commutative nilpotent algebras of dimension 5 of class 3 described in Section 9 of Chapter 2. Regular representations of these algebras are as follows:

$$N_1 = [e_{21} + e_{52}, e_{31} + e_{53}, e_{41} + e_{54} + e_{64}, e_{51}, e_{61}],$$

$$N_2 = [e_{21} + e_{52}, e_{31} + e_{53} + 2e_{63}, e_{41} + e_{54} + e_{64}, e_{51}, e_{61}],$$

$$N_3 = [e_{21} + e_{52}, e_{31} + e_{53} + ie_{63} + e_{64}, e_{41} + e_{63} + e_{54} - ie_{64},$$
$$e_{51}, e_{61}],$$

$$N_4 = [e_{21} + e_{52} + e_{62}, e_{31} + e_{53} + ie_{63} + e_{64},$$
$$e_{41} + e_{63} + e_{54} - ie_{64}, e_{51}, e_{61}],$$

$$N_5 = [e_{21} + e_{52} + e_{63}, e_{31} + e_{62} + e_{53} - ie_{64},$$
$$e_{41} - ie_{63} + e_{54}, e_{51}, e_{61}],$$

$$N_6 = [e_{21} + e_{52} + e_{64}, e_{31} + e_{53} + ie_{64}, e_{41} + e_{62} + ie_{63}, e_{51}, e_{61}],$$

$$N_7 = [e_{21} + e_{62}, e_{31} + e_{63}, e_{41} + e_{64}, e_{51} + e_{65}, e_{61}],$$

$$N_8 = [e_{21} + e_{42} + e_{53}, e_{31} + e_{52} + e_{63}, e_{41}, e_{51}, e_{61}],$$

$$N_9 = [e_{21} + e_{32}, e_{31}, e_{41}, e_{51}, e_{61}],$$

$$N_{10} = [e_{21} + e_{42}, e_{31} + e_{43}, e_{41}, e_{51}, e_{61}],$$

$$N_{11} = [e_{21} + e_{52}, e_{31} + e_{53}, e_{41} + e_{54}, e_{51}, e_{61}],$$

$$N_{12} = [e_{21} + e_{42}, e_{31} + e_{43} + e_{53}, e_{41}, e_{51}, e_{61}],$$

$$N_{13} = [e_{21} + e_{42} + ie_{52} + e_{62}, e_{31} + e_{52} + e_{43}$$
$$- ie_{53}, e_{41}, e_{51}, e_{61}].$$

For what follows, it is important to note that out of all the algebras N_1–N_{13} enumerated, only the algebra N_7 has signature $(1, 4, 1)$.

Subcase (b) $\nu = 2$, $m = 1$, $\mu = 3$. Here, the matrices g of the algebra N can be reduced simultaneously to the form

$$\begin{bmatrix} 0_{22} & 0_{21} & 0_{23} \\ a(g) & 0 & 0_{13} \\ c(g) & b(g) & 0_{33} \end{bmatrix},$$

where

$$a(g) = [\alpha_1(g)\alpha_{12}(g)], \qquad b(g) = \begin{bmatrix} \beta_1(g) \\ \beta_2(g) \\ \beta_3(g) \end{bmatrix},$$

and $c(g)$ is an arbitrary 3×2 matrix. If the matrix

$$u = \begin{bmatrix} 0_{22} & 0_{21} & 0_{23} \\ a & 0 & 0_{13} \\ 0_{32} & b & 0_{33} \end{bmatrix} \in N$$

does not belong to the annihilator of the algebra N, then $a \neq 0$ and $b \neq 0$. Obviously, there exist matrices $s \in GL(2, P)$ and $t \in GL(3, P)$ such that $as = [1, 0] = a_1$ and

$$t^{-1}b = \begin{bmatrix} 1 \\ 0 \\ 0 \end{bmatrix} = b_1.$$

Now, if $r = [s, 1, t]$, then $r^{-1}ur = e_{31} + e_{43}$. Consequently, we may assume that $e_{31} + e_{43}$ belongs to N. From the condition $(e_{31} + e_{43})g = g(e_{31} + e_{43})$, we find

$$g = \lambda(e_{31} + e_{43}) + g_0,$$

where

$$g_0 = \begin{bmatrix} 0_{32} & 0_{34} \\ c & 0_{34} \end{bmatrix}.$$

Thus, in subcase (b), we obtain the single algebra

$$N_{14} = [e_{31} + e_{43}, e_{41}, e_{42}, e_{51}, e_{52}, e_{61}, e_{62}].$$

Subcase (c) $v = m = \mu = 2$. Here, the matrices of the algebra N can be reduced simultaneously to the form

$$g = \begin{bmatrix} 0 & 0 & 0 \\ a(g) & 0 & 0 \\ c(g) & b(g) & 0 \end{bmatrix},$$

where 0, $a(g)$, and $b(g)$ are 2×2 matrices and 0 is the zero matrix. We need to consider four possibilities:

(1) N contains a matrix of rank 4; (2) N contains no matrix of rank 4 but it contains a matrix g such that $a(g)$ is of rank 2; (3) N contains no matrix g such that $a(g)$ is of rank 2 but it contains a matrix g such that $b(g)$ is of rank 2; (4) $a(g)$ and $b(g)$ are singular for every matrix $g \in N$. We take up these four possibilities separately:

(1) N contains a matrix g of rank 4. Then, $a(g)$ and $b(g)$ are nonsingular matrices. Therefore, there exist s_1 and $s_3 \in GL(2, P)$ such that $as_1 = s_3^{-1}b = E_2$. Now, if $t = [s_1, E_2, s_3]$, then

$$t^{-1}gt = \begin{bmatrix} 0 & 0 & 0 \\ E_2 & 0 & 0 \\ * & E_2 & 0 \end{bmatrix}.$$

Consequently, we may assume that N contains a matrix g_1 such that $a(g_1) = b(g_1) = E_2$ and $c(g_1) = 0$. From the condition $gg_1 = g_1g$, we obtain

$$g = \begin{bmatrix} 0 & 0 & 0 \\ a(g) & 0 & 0 \\ c(g) & a(g) & 0 \end{bmatrix},$$

where $a(g)$ ranges over a maximal commutative subalgebra of P_2 as g ranges over N. P_2 contains two maximal commutative subalgebras $A_1 = [e_{11}, e_{22}]$ and $A_2 = [E_2, e_{21}]$. Consequently, this case produces two algebras:

$$N_{15} = [e_{31} + e_{53}, e_{42} + e_{64}, e_{51}, e_{52}, e_{61}, e_{62}],$$
$$N_{16} = [e_{31} + e_{42} + e_{53} + e_{64}, e_{41} + e_{63}, e_{51}, e_{52}, e_{61}, e_{62}].$$

(2) Suppose now that N contains no matrix of rank 4 but it does contain a matrix g_1 such that the rank of $a(g_1)$ is 2. Then, we may assume that

$$g_1 = \begin{bmatrix} 0 & 0 & 0 \\ E_2 & 0 & 0 \\ 0 & b_1 & 0 \end{bmatrix},$$

where $b_1 = [1, 0]$. Then, the matrix

$$g_2 = \begin{bmatrix} 0 & 0 & 0 \\ a_2 & 0 & 0 \\ 0 & b_2 & 0 \end{bmatrix},$$

where

$$a_2 = \begin{bmatrix} 0 & \alpha \\ \beta & \gamma \end{bmatrix},$$

belongs to N and is not the zero matrix. Then, from the condition $g_2 g_1 = g_1 g_2$, we obtain

$$b_2 = \begin{bmatrix} 0 & \alpha \\ 0 & 0 \end{bmatrix}$$

where $\alpha \neq 0$ since, in the opposite case, we would have $g_2 = 0$ from Kravchuk's third theorem. Consequently, we may assume that $\alpha = 1$. Let us set

$$t_1 = \begin{bmatrix} 1 & 0 \\ 2^{-1}\delta & 1 \end{bmatrix},$$

where $\delta = \gamma + (\gamma^2 + 4\beta)^{1/2}$. Then, $b_1 t_1 = b_1$ and

$$t_1^{-1} a_2 t_1 = 2^{-1}\delta E_2 + \begin{bmatrix} 0 & 1 \\ 0 & \gamma - \delta \end{bmatrix}.$$

Now, if $t = [t_1, t_1, E_2] \in GL(6, P)$, then $t^{-1} g_1 t = g_1$ and $t^{-1} g_2 t = 2^{-1}\delta g_1 + g_3$, where

$$g_3 = \begin{bmatrix} 0 & 0 & 0 \\ a_3 & 0 & 0 \\ 0 & b_3 & 0 \end{bmatrix}, \qquad a_3 = \begin{bmatrix} 0 & 1 \\ 0 & \gamma - \delta \end{bmatrix}, \qquad b_3 = \begin{bmatrix} 0 & 1 \\ 0 & 0 \end{bmatrix}.$$

Thus, we may assume that N contains the matrices g_1 and g_3.

If $\gamma = \delta$, we obviously obtain the algebra

$$N_{17} = [e_{31} + e_{42} + e_{53}, e_{32} + e_{54}, e_{51}, e_{52}, e_{61}, e_{62}].$$

If $\gamma \neq \delta$, we set

$$t_2 = \begin{bmatrix} 1 & 0 \\ \gamma - \delta & \gamma - \delta \end{bmatrix}.$$

Then,

$$t_2^{-1} a_3 t_2 = (\gamma - \delta)\left(E_2 + \begin{bmatrix} 0 & 1 \\ 0 & -1 \end{bmatrix} \right),$$

and $b_1 t_2 = b_1$. Define

$$a_4 = \begin{bmatrix} 0 & 1 \\ 0 & -1 \end{bmatrix}, \qquad \text{and} \qquad d = [t_2, t_2, E_2].$$

Then, $d^{-1}g_1 d = g_1$ and $d^{-1}g_3 d = (\gamma - \delta)(g_1 + g_4)$, where

$$\begin{bmatrix} 0 & 0 & 0 \\ a_4 & 0 & 0 \\ 0 & b_4 & 0 \end{bmatrix} = g_4,$$

with

$$b_4 = \begin{bmatrix} 0 & 1 \\ 0 & 0 \end{bmatrix}.$$

Obviously, $g_1 = e_{31} + e_{42} + e_{53}$ and $g_4 = e_{32} - e_{42} + e_{54}$. Consequently, for $\gamma \neq \delta$, we obtain the algebra

$$N_{13} = [e_{31} + e_{42} + e_{53}, e_{32} - e_{42} + e_{54}, e_{51}, e_{52}, e_{61}, e_{62}].$$

It remains to show that N_{17} and N_{18} are nonconjugate in P_6. Suppose that $GL(6, P)$ is a matrix r such that $r^{-1} N_{17} r = N_{18}$. Now, if M is the annihilator of N_{17}, then $r^{-1} M r = M$. From this it follows that

$$r = \begin{bmatrix} r_1 & 0 & 0 \\ * & r_2 & 0 \\ * & * & r_3 \end{bmatrix},$$

where 0, r_i (for $i = 1, 2, 3$), and $*$ are 2×2 matrices. As one can easily show, for g_1 and g_3 in N_{17},

$$r^{-1}g_1 r = r^{-1} \begin{bmatrix} 0 & 0 & 0 \\ E_2 & 0 & 0 \\ 0 & b_1 & 0 \end{bmatrix}, \qquad r = \begin{bmatrix} 0 & 0 & 0 \\ r_2^{-1} r_1 & 0 & 0 \\ * & * & 0 \end{bmatrix} \in N_{18}.$$

$$(3.63)$$

$$r^{-1}g_3r = r^{-1}\begin{bmatrix} 0 & 0 & 0 \\ a_3 & 0 & 0 \\ 0 & b_3 & 0 \end{bmatrix}, \qquad r = \begin{bmatrix} 0 & 0 & 0 \\ r_2^{-1}a_3r_1 & 0 & 0 \\ * & * & 0 \end{bmatrix} \in N_{18},$$

$$(3.64)$$

where

$$a_3 = \begin{bmatrix} 0 & 1 \\ 0 & 0 \end{bmatrix}.$$

From formulas (3.63) and (3.64), we obtain

$$r_2^{-1}r_1 = \lambda E_2 + \mu \begin{bmatrix} 0 & 1 \\ 0 & -1 \end{bmatrix} = \begin{bmatrix} \lambda & \mu \\ 0 & \lambda - \mu \end{bmatrix}, \qquad (3.65)$$

$$r_2^{-1}\begin{bmatrix} 0 & 1 \\ 0 & 0 \end{bmatrix}r_1 = \lambda_1 E_2 + \mu_1 \begin{bmatrix} 0 & 1 \\ 0 & -1 \end{bmatrix} = \begin{bmatrix} \lambda_1 & \mu_1 \\ 0 & \lambda_1 - \mu_1 \end{bmatrix}, \quad (3.66)$$

where $\lambda, \mu, \lambda_1, \mu_1 \in P$. Obviously, $\lambda_1(\lambda_1 - \mu_1) = 0$. Consequently, either $\lambda_1 = 0$ and $\mu_1 \neq 0$ or $\lambda_1 = \mu_1 \neq 0$. Let us suppose first that $\lambda_1 = 0$. Then, if we set

$$r_1 = \begin{bmatrix} \rho_{11} & \rho_{12} \\ \rho_{21} & \rho_{22} \end{bmatrix} \qquad \text{and} \qquad r_2 = \begin{bmatrix} \tau_{11} & \tau_{12} \\ \tau_{21} & \tau_{22} \end{bmatrix}$$

in (3.66), we obtain

$$\begin{bmatrix} 0 & 1 \\ 0 & 0 \end{bmatrix}\begin{bmatrix} \rho_{11} & \rho_{12} \\ \rho_{21} & \rho_{22} \end{bmatrix} = \begin{bmatrix} \tau_{11} & \tau_{12} \\ \tau_{21} & \tau_{22} \end{bmatrix}\begin{bmatrix} 0 & \mu_1 \\ 0 & -\mu_1 \end{bmatrix}.$$

From this, we obtain

$$\begin{bmatrix} \rho_{21} & \rho_{22} \\ 0 & 0 \end{bmatrix} = \begin{bmatrix} 0 & \mu_1(\tau_{11} - \tau_{12}) \\ 0 & \mu_1(\tau_{21} - \tau_{22}) \end{bmatrix}, \qquad \rho_{21} = 0, \qquad \tau_{21} = \tau_{22}.$$

$$(3.67)$$

From (3.65) and the equation $\rho_{21} = 0$, it follows that $\tau_{21} = 0$. On the basis of (3.67), r_2 is a singular matrix, which is impossible. Suppose now that $\lambda_1 = \mu_1 \neq 0$. Then, from (3.61), we obtain

$$\begin{bmatrix} \rho_{21} & \rho_{22} \\ 0 & 0 \end{bmatrix} = \begin{bmatrix} \tau_{11} & \tau_{12} \\ \tau_{21} & \tau_{22} \end{bmatrix} \begin{bmatrix} \lambda_1 & \lambda_1 \\ 0 & 0 \end{bmatrix} = \begin{bmatrix} \tau_{11}\lambda_1 & \tau_{11}\lambda_1 \\ \tau_{21}\lambda_1 & \tau_{21}\lambda_1 \end{bmatrix}. \quad (3.67a)$$

Consequently, $\tau_{21} = 0$. But then, $\rho_{21} = 0$ and, by virtue of (3.67a), $\tau_{11} = 0$; that is, r_2 is a singular matrix. Thus, the algebras N_{17} and N_{18} are nonconjugate in P_6.

(3) Suppose now that N contains no matrix g such that $a(g)$ is equal to 2 but that it does contain a matrix g_1 such that the rank of $b(g_1)$ is 2. Then, let us construct the algebra

$$L = d^{-1}N'd, \quad (3.68)$$

where N' is obtained from N by transposing all its matrices and

$$d = \begin{bmatrix} 0 & 0 & E_2 \\ 0 & E_2 & 0 \\ E_2 & 0 & 0 \end{bmatrix}.$$

As one can easily show, the matrices of the algebra L are of the form

$$h = \begin{bmatrix} 0 & 0 & 0 \\ b'(g) & 0 & 0 \\ c'(g) & a'(g) & 0 \end{bmatrix}, \quad (3.69)$$

where

$$g = \begin{bmatrix} 0 & 0 & 0 \\ a(g) & 0 & 0 \\ c(g) & b(g) & 0 \end{bmatrix}$$

ranges over the algebra N and where $a'(g)$ and $b'(g)$ are the transposes of $a(g)$ and $b(g)$, respectively.

On the basis of formula (3.69), L is conjugate with one of the algebras of case (2); that is, we may assume that L is either the algebra N_{17} or the algebra N_{18}. From (3.68), we now see that $N = d^{-1}L'd = d^{-1}N'_{17}d$ or $N = d^{-1}N'_{18}d$. Thus, case (3) yields two algebras:

$$N_{19} = N'_{17} = [e_{13} + e_{24} + e_{35}, e_{23} + e_{45}, e_{15}, e_{25}, e_{16}, e_{26}],$$

$$N_{20} = N'_{18} = [e_{13} + e_{24} + e_{35}, e_{23} - e_{24} + e_{45}, e_{15}, e_{25}, e_{16}, e_{26}].$$

(4) Suppose, finally, that the matrices $a(g)$ and $b(g)$ are singular for every $g \in N$. Let g_1 denote a matrix in N that does not belong to the annihilator of N. Then, $a(g_1)$ and $b(g_1)$ both have rank 1. Consequently, we may assume that

$$g_1 = \begin{bmatrix} 0 & 0 & 0 \\ a_1 & 0 & 0 \\ 0 & b_1 & 0 \end{bmatrix},$$

where $a_1 = [1, 0] = b_1$. Suppose that $g_2 \in N$ and

$$g_2 = \begin{bmatrix} 0 & 0 & 0 \\ a_2 & 0 & 0 \\ 0 & b_2 & 0 \end{bmatrix}.$$

From the condition $g_1g_2 = g_2g_1$, we find

$$b_1a_2 = b_2a_1. \tag{3.70}$$

If

$$a_2 = \begin{bmatrix} \alpha & \beta \\ \gamma & \delta \end{bmatrix} \quad \text{and} \quad b_2 = \begin{bmatrix} \alpha_1 & \beta_1 \\ \gamma_1 & \delta_1 \end{bmatrix},$$

then, from (3.70), we obtain

$$\begin{bmatrix} \alpha & \beta \\ 0 & 0 \end{bmatrix} = \begin{bmatrix} \alpha_1 & 0 \\ \gamma_1 & 0 \end{bmatrix}.$$

From this we get $\alpha_1 = \alpha$, $\beta = 0$, and $\gamma_1 = 0$. Consequently,

$$a_2 = \begin{bmatrix} \alpha & 0 \\ \gamma & \delta \end{bmatrix} \quad \text{and} \quad b_2 = \begin{bmatrix} \alpha & \beta_1 \\ 0 & \delta_1 \end{bmatrix}.$$

By hypothesis, $\delta = \delta_1 = 0$. Consequently, if g_1 and g_2 are linearly independent, we may assume that

$$a_2 = \begin{bmatrix} 0 & 0 \\ 1 & 0 \end{bmatrix} \quad \text{and} \quad b_2 = \begin{bmatrix} 0 & \beta \\ 0 & 0 \end{bmatrix},$$

where $\beta \neq 0$. If $t_1 = [1, \beta^{-1}]$, then $b_1 t_1 = b_1$ and

$$b_2 t_1 = \begin{bmatrix} 0 & 1 \\ 0 & 0 \end{bmatrix}.$$

Suppose that $t = [E_4, t_1]$. Then, $t^{-1} g_1 t = g_1$ and

$$t^{-1} g_2 t = g_3 = \begin{bmatrix} 0 & 0 & 0 \\ a_2 & 0 & 0 \\ 0 & b_3 & 0 \end{bmatrix},$$

where

$$a_2 = \begin{bmatrix} 0 & 0 \\ 1 & 0 \end{bmatrix} \quad \text{and} \quad b_3 = \begin{bmatrix} 0 & 1 \\ 0 & 0 \end{bmatrix}.$$

Since, for arbitrary $g \in N$, the rank of $a(g)$ does not exceed 1, it follows that

$$N = [g_1, g_3, e_{51}, e_{52}, e_{61}, e_{62}].$$

Consequently, in case (4), we obtain the algebra

$$N_{21} = [e_{31} + e_{51}, e_{41} + e_{54}, e_{51}, e_{52}, e_{61}, e_{62}].$$

Subcase (d), $\nu \neq 1$, $\mu = 1$. Here, we may, in accordance with Theorem 9 of the preceding chapter, assume that N is obtained by transposing an algebra with signature $(i, 5 - \gamma, \gamma)$, where $\gamma \neq 1$. Consequently, by transposing the algebras N_1–N_6 and N_8–N_{13}, we obtain

$$N_{22} = N_1{}' = [e_{12} + e_{25}, e_{13} + e_{35}, e_{14} + e_{45} + e_{46}, e_{15}, e_{16}],$$

$$N_{23} = N_2{}' = [e_{12} + e_{25}, e_{13} + e_{35} + 2e_{36}, e_{14} + e_{45} + e_{46},$$
$$e_{15}, e_{16}],$$

$$N_{24} = N_3{}' = [e_{12} + e_{25}, e_{13} + e_{35} + ie_{36} + e_{46}, e_{14}$$
$$+ e_{36} + e_{45} - ie_{46}, e_{15}, e_{16}],$$

$$N_{25} = N_4{}' = [e_{12} + e_{25} + e_{26}, e_{13} + e_{35} + ie_{36} + e_{46}, e_{14}$$
$$+ e_{36} + e_{45} - ie_{46}, e_{15}, e_{16}],$$

$$N_{26} = N_5{}' = [e_{12} + e_{25} + e_{36}, e_{13} + e_{26} + e_{35} - ie_{46}, e_{14}$$
$$- ie_{36} + e_{45}, e_{15}, e_{16}],$$

$$N_{27} = N_6{}' = [e_{12} + e_{25} + e_{46}, e_{13} + e_{35} + ie_{46}, e_{14}$$
$$+ e_{26} + ie_{36}, e_{15}, e_{16}],$$

$$N_{28} = N_8{}' = [e_{12} + e_{24} + e_{35}, e_{13} + e_{25} + e_{36}, e_{14}, e_{15}, e_{16}],$$

$$N_{29} = N_9{}' = [e_{12} + e_{23}, e_{13}, e_{14}, e_{15}, e_{16}],$$

$$N_{30} = N_{10}{}' = [e_{12} + e_{24}, e_{13} + e_{34}, e_{14}, e_{15}, e_{16}],$$

$$N_{31} = N_{11}{}' = [e_{12} + e_{25}, e_{13} + e_{35}, e_{14} + e_{45}, e_{15}, e_{16}],$$

$$N_{32} = N_{12}' = [e_{12} + e_{24}, e_{13} + e_{34} + e_{35}, e_{14}, e_{15}, e_{16}],$$

$$N_{33} = N_{13}' = [e_{12} + e_{24} + ie_{25} + e_{26}, e_{13} + e_{25}$$

$$+ e_{34} - ie_{35}, e_{14}, e_{15}, e_{16}].$$

Subcase (e), $\nu = 3$, $m = 1$, $\mu = 2$.   On the basis of Theorem 8 of the preceding chapter, the algebra N in the present subcase is conjugate with the algebra N_{14}', where N_{14} is the algebra, mentioned above, with signature $(2, 1, 3)$. Consequently, we obtain the algebra

$$N_{34} = [e_{13} + e_{34}, e_{14}, e_{24}, e_{15}, e_{25}, e_{16}, e_{26}].$$

This completes our study of the case $l = 3$ (with $n = 6$).

For $l = 2$, we obtain, in accordance with Section 3 of Chapter 2, the five algebras

$$N_{35} = [e_{21}, e_{31}, e_{41}, e_{51}, e_{61}],$$

$$N_{36} = [e_{31}, e_{32}, e_{41}, e_{42}, e_{51}, e_{52}, e_{61}, e_{62}],$$

$$N_{37} = [e_{41}, e_{42}, e_{43}, e_{51}, e_{52}, e_{53}, e_{61}, e_{62}, e_{63}],$$

$$N_{38} = [e_{51}, e_{52}, e_{53}, e_{54}, e_{61}, e_{62}, e_{63}, e_{64}],$$

$$N_{39} = [e_{61}, e_{62}, e_{63}, e_{64}, e_{65}].$$

For $l = 4$, we obtain, in accordance with Theorem 2, the following fourteen algebras:

$$N_{40} = [a, a^2, a^3, e_{65}, e_{61}, e_{45}],$$

where

$$a = e_{21} + e_{32} + e_{43}; \tag{3.71}$$

$$N_{41} = [a, a^2, a^3, e_{51} + e_{62} + e_{35} + e_{46}, e_{61} + e_{45}],$$

where

$$a = e_{21} + e_{32} + e_{43} + e_{65}; \tag{3.72}$$

$$N_{42} = [a, a^2, a^3, e_{51} + e_{62} + e_{35} + e_{46} + 2ie_{65}, e_{61} + e_{45}],$$

where a is the matrix (3.72);

$$N_{43} = [a, a^2, a^3, e_{61} + e_{35} + e_{46}, e_{45}],$$

where a is the matrix (3.72);

$$N_{44} = [a, a^2, a^3, e_{35} + e_{46} + e_{65}, e_{45}],$$

where a is the matrix (3.72);

$$N_{45} = [a, a^2, a^3, e_{35} + e_{46}, e_{45}],$$

where a is the matrix (3.72);

$$N_{46} = [b, b^2, b^3, e_{16} + e_{53} + e_{64}, e_{54}];$$
$$N_{47} = [b, b^2, b^3, e_{53} + e_{64} + e_{56}, e_{54}];$$
$$N_{48} = [b, b^2, b^3, e_{53} + e_{64}, e_{54}],$$

where $b = e_{12} + e_{23} + e_{34} + e_{56}$;

$$N_{49} = [a, a^2, a^3, e_{51}, e_{56}];$$
$$N_{50} = [a, a^2, a^3, e_{51} + e_{45}, e_{61}],$$

where a is the matrix (3.71);

$$N_{51} = [c, c^2, c^3, e_{15}, e_{16}];$$
$$N_{52} = [c, c^2, c^3, e_{15} + e_{54}, e_{16}],$$

where $c = e_{12} + e_{23} + e_{34}$;

$$N_{53} = [a, a^2, a^3, e_{51} + e_{45}, e_{61} + e_{46}],$$

where a is the matrix (3.71).

For $l = 5$, we obtain, in accordance with Theorem 1, the three algebras

$$N_{54} = [a, a^2, a^3, a^4, e_{61}],$$
$$N_{55} = [a, a^2, a^3, a^4, e_{56}],$$
$$N_{56} = [a, a^2, a^3, a^4, e_{61} + e_{56}],$$

where

$$a = e_{21} + e_{32} + e_{43}, e_{54}.$$

For $l = 6$,

$$N_{57} = [a, a^2, a^3, a^4, a^5],$$

where

$$a = e_{21} + e_{32} + e_{43} + e_{54} + e_{65}.$$

$$* \quad * \quad *$$

Let us now pose some problems.

(1) Let C denote the field of complex numbers, let C_n denote the algebra of all $n \times n$ matrices over C, let $U_{n,k}$ denote the set of all maximal commutative nilpotent sub-algebras of C_n with index of nilpotency k, and let $\bar{U}_{n,k}$ denote the set of classes of algebras in $U_{n,k}$ that are conjugate in C_n. For what pairs (n, k) is the set $\bar{U}_{n,k}$ finite?

(2) Indicate methods of constructing families of algebras $M_1, \ldots, M_r$ possessing the following properties:

(i) M_j is the family of maximal commutative subalgebras of C_n that depends on finitely many parameters.

(ii) Every maximal commutative subalgebra of C_n is conjugate in C_n with some subalgebra of some family M_j.

(3) What is the number of essential parameters of the set of conjugacy classes of maximal commutative subalgebras of the algebra C_n?

(4) To what extent can the results of Chapter 3 be carried over to the case of a real base field?

(5) Let P denote a finite field with characteristic p and let Γ denote the Sylow p-subgroup of $GL(n, P)$. What is the structure of the maximal commutative subgroups of the group Γ?

(6) Let Δ denote an arbitrary field. What is the minimum dimension $r(n, \Delta)$ of a maximal commutative subalgebra of the full matrix algebra Δ_n? Does $r(n, \Delta)$ depend on the field Δ?

(7) What is the structure of a commutative indecomposable algebra with two irreducible parts over a perfect field?

Bibliography

1. Büke, A., Untersuchungen über kommutativ-assoziative und nilpotente Algebren von Index 3 und von der Charakteristik 2, *Istanbul Univ. Fen Fak. Mec.* **A19**, Suppl. 145 S (1954).

2. Bogdanov, Yu. S., and Chebotarev, G. N., O matritsakh, kommutiruyushchikh so svoyey proizvodnoy (On matrices that commute with their derivatives), *Izv. vuzov Matem.* No. 4, 27–37 (1959).

3. van der Waerden, B. L., "Die gruppentheoretische Methode in der Quantenmechanik." Springer, Berlin, 1932.

4. van der Waerden, B. L., "Algebra," 6th ed., Springer, Berlin, 1964. [English transl., "Modern Algebra," 2nd ed. Ungar, New York, 1949—50.]

5. Wan Zhe-xian, Li Gen-dao, The two theorems of Schur on commutative matrices, *Chinese Math.* **5**, No. 1, 156–164 (1964).

6. Weyl, H., "The Classical Groups: their Invariants and Representations." Princeton Univ. Press, Princeton, New Jersey, 1939.

7. Gantmakher, F. R., "Teoriya Matrits." [English transl. "The Theory of Matrices" Chelsea, New York, 1959.]

8. Gauthier, Luc., Commutation des matrices et congruences d'ordre un., *Bull. Soc. Math. France* **84**, 283–294 (1956).

9. Gerstenhaber, M., On dominance and varieties of commuting matrices, *Ann. of Math.* **73**, No. 2 (1961).

10. Dyment, Z. M., Maksimal'nyye kommutativnyye nil'potentnyye podalgebry matrichnoy algebry shestoy stepeni (Maximal commutative nilpotent subalgebras of a 6×6 matrix algebra), *Izv. Akad. Nauk* Belorussk. SSR, *Ser. FIZ.-MAT.* No. 4 (1965).

11. Jacobson, N., Schur's theorems on commutative matrices, *Bull. Amer. Math. Soc.* **50**, No. 6 (1944).

12. Jacobson, N., "*The Theory of Rings*," Amer. Math. Soc., Providence, Rhode Island, 1943.

13. Dzazin, M. P., Dungey, J. W., and Gruenberg, K. W., Some theorems on commutative matrices, *J. London Math. Soc.* **26**, 221–228 (1951).

14. Dubisch, K., and Perlis, S., On total nilpotent algebras, *Amer. J. Math.* **73**, 439–452 (1951).

15. Jordan, C., Réduction d'un réseau de formes quadratiques ou bilinéaires, *J. Math. Pures Appl.* 403–438 (1906); 3–51 (1907).

16. Jordan, C., Groupes abéliens généraux contenues dans les groupes linéaires à moins de sept variables, *J. Math. Pures Appl.* **3**, 213–266 (1907).

17. Egan, M. F., and Ingram, R. E., On commutative matrices, *Math. Gaz.* **37**, No. 320, 107–110 (1953).

18. Kravchuk, M. F., O gruppakh perestanovochnykh matrits (On groups of commutative matrices), *Soobshch. Khar'kovskogo Mat. Obshch-va* Ser. 2, **14**, No. 4, 163–176 (1914).

19. Kravchuk, M. F., Do teorii pereminnikh matryts (On the theory of variable matrices), *Ukr. Akad. Nauk Zap. Fiz.-Mat. Viddilu* **2**, 28–33 (1924).

20. Kravchuk, M. F., Pereminni mnozhyny liniinykh peretvoren (Variable sets of linear transformations), *Zap. KSGI* **1**, 23–58 (1956).

21. Krawtschouk, M., Über vertauschbare Matrizen, *Rend. Circ. Mat. Palermo* **51**, 126–130 (1927).

22. Kravchuk, M. F., O strukture perestanovochnykh grupp matrits (On the structure of commutative groups of matrices), *Tr. Second All-Union Mathematical Congress*, **2**, 11–12 (1934).

23. Kravchuk, M. F., and Gol'dbaum, Ya. S., Pro grupi kommutativnykh matryts (On groups of commutative matrices), *Tr. Kyivs'kogo aviatsyonnogo ynstytuta* **5**, 12–23 (1936).

24. Courter, R. C., The dimension of maximal commutative subalgebras of K_n, *Duke Math. J.* **32**, No. 2 (1965).

25. Mal'tsev, A. I., Kommutativnyye podalgebry poluprostykh algebry Li (Commutative subalgebras of semisimple Lie algebras), *Izv. Akad. Nauk SSSR, Ser. Mat.* **9**, 291–300 (1943).

26. Mal'tsev, A. I., "Osnovy lineynoy algebry" (Fundamentals of linear algebra), Moscow and Leningrad, 1956.

27. Morozov, V. V., O kommutativnykh matritsakh (On commutative matrices), *Uch. Zap. Kazanskogo Gos. Univ.-ta* **112**, Book 9, *Sb. Rabot NII Mat. Mekh. im. N. G. Chebotareva* (1952).

28. Suprunenko, D. A., Razreshinyye i nil'potentnyye lineynyye gruppy (Solvable and nilpotent linear groups), Dissertation, Institut im. V. A. Steklova (1955).

29. Suprunenko, D. A., O maksimal'nykh kommutativnykh podalgebrakh polnoy lineynoy algebry (On maximal commutative subalgebras of a complete linear algebra), *Uspekhi Mat. Nauk* **2**, No. 3 (69), 181–184 (1956).

30. Suprunenko, D. A., Maksimal'nyye kommutativnyye nil'potent-nyye podalgebry klassa $n - 2$ polnoy matrichnoy algebry (Maximal commutative nilpotent subalgebras of class $n - 2$ of a complete matrix algebra), *Vestsi Akad. Nauk Belorussk. SSR. Ser. Fiz.-Tekhn.* No. 3, 135–145 (1956).

31. Suprunenko, D. A., and Tyshkevich, R. I., Privodimyye lokal'no nil'potentnyye lineynyye gruppy (Reducible locally nilpotent linear groups), *Izv. Akad. Nauk, Belorussk. SSR, Ser.-Mat.* **24**, 787–806 (1960).

32. Suprunenko, D. A., Usloviya polnoy privodimosti razreshimoi lineynoy gruppy (Conditions for complete reducibility of a solvable linear group), *Dokl. Akad. Nauk Belorussk. SSR* **5**, No. 8, 321–323 (1961).

33. Suprunenko, D. A., O privodimykh matrichnykh gruppakh (On reducible matrix groups), *Dokl. Akad. Nauk Belorussk. SSR* **5**, No. 9, 371–374 (1961).

34. Suprunenko, D. A., O maksimal'nykh kommutativnykh matrich-nykh algebrakh i maksimal'nykh kommutativnykh matrichnykh gruppakh (On maximal commutative matrix algebras and maximal commutative matrix groups) *Dokl. Akad. Nauk Belorussk. SSR* **8**, No. 7, 425–428 (1964).

35. Taussky, O., Commutativity in finite matrices, *Amer. Math. Monthly* **64**, No. 4, 229–235 (1957).

36. Feit, W. and Fine, N. J., Pairs of commuting matrices over a finite field, *Duke Math. J.* **27**, No. 1, 91–94 (1960).

37. Phillips, H. B., Functions of Matrices, *Amer. J. Math.* 266–278 (1919).

38. Frobenius, G., Über vertauschbare Matrizen, *Sitzber. Berlin Akad.* 601–614 (1896).

39. Hamburger, H. L., A theorem on commutative matrices, *J. London Math. Soc.* **24**, 200–206 (1949).

40. Charles, B., Sur la permutabilité d'un opérateur linéaire, *Compt. Rend. Acad. Sci.* **236**, 1722 (1953).

41. Charles, B., Un critère de maximalité pour les anneaux com-mutatifs d'opérateurs linéaires, *Compt. Rend. Acad. Sci.* **236**, 1835–1837 (1953).

42. Charles, B., Un exemple general d'anneau commutatif d'opéra-teurs linéaires tel que $R'' \neq r(R, i)$, *Compt. Rend. Acad. Sci.* **236**, 2027–2029 (1953).

43. Charles, B., Sur algèbres des opérateurs linéaires, *J. Math. Pures Appl.* **33**, 2, 81–145 (1954).

44. Schur, I., Zur Theorie der vertauschbaren Matrizen, 1, *J. Reine Angew. Math.* **130**, 66–76 (1905).

45. Robinson, D. W., On matrix commutators of higher order, *Canad. J. Math.* **17**, No. 4, 527–532 (1965).

Subject Index

A

Abelian group, 13
Algebra
  associative, 9
    theory of, 9
  decomposable, 36
  finite-dimensional, 6, 9
  indecomposable, 36
  linear, 25
  pairwise-nonisomorphic, 121
  semisimple, 9
Annihilator, defined, 51

B

Bilinear forms, 68, 69

C

Centralizer, defined, 5
Commutative algebra of matrices, 22
Commutative matrices, 1
  normal forms of, 9
  properties of, 1
Commutative matrix algebras
  low-order, 132
  nilpotent algebras
    dimension of, 88
    regular representation of, 60
Commutative set, 19
  subalgebras, 35
    general properties of, 41
    problem of constructing
      maximal, 45
Complement, invariant, 3
Complex-number field, 98

D

Dedekind's proposition, 9
Diagonal matrices, 78
Diagonals, 38
Dyment, Z. M., 72

F

Family of algebras, one-parameter, 106
Frobenius' proposition, 10

I

Identity mapping, 39
Invariant complement, 3

J

Jordan normal form, 33, 34, 51, 98, 101

K

Kravchuk signature, 53, 65, 105, 136
Kravchuk's first theorem, 51, 62
  normal form, 51, 53, 55, 60, 89, 99
  second theorem, 53
  third theorem, 55, 57, 60, 63, 64, 89, 141

L

Linear algebra, 25
  operators
    irreducible, 2
    reducible, 2
  transformation, nonsingular, 126